Intelligent Prognostics for Engineering Systems with Machine Learning Techniques

The text discusses the latest data-driven, physics-based, and hybrid approaches employed in each stage of industrial prognostics and reliability estimation.

The book

- Discusses basic as well as advance research in the field of prognostics.
- Explores integration of data collection, fault detection, degradation modeling, and reliability prediction in one volume.
- Covers prognostics and health management (PHM) of engineering systems.
- Discusses latest approaches in the field of prognostics based on machine learning.

The text deals with tools and techniques used to predict/extrapolate/forecast the process behavior, based on current health state assessment and future operating conditions with the help of machine learning. It will serve as a useful reference text for senior undergraduate, graduate students, and academic researchers in areas such as industrial and production engineering, manufacturing science, electrical engineering, and computer science.

Advanced Research in Reliability and System Assurance Engineering

Series Editor: Mangey Ram, Professor, Graphic Era University, Uttarakhand, India

Stochastic Models in Reliability Engineering
Lirong Cui, Ilia Frenkel, and Anatoly Lisnianski

Predictive Analytics
Modeling and Optimization
Vijay Kumar and Mangey Ram

Design of Mechanical Systems Based on Statistics
A Guide to Improving Product Reliability
Seong-woo Woo

Social Networks
Modeling and Analysis
Niyati Aggrawal and Adarsh Anand

Operations Research
Methods, Techniques, and Advancements
Edited by Amit Kumar and Mangey Ram

Statistical Modeling of Reliability Structures and Industrial Processes
Edited by Ioannis S. Triantafyllou and Mangey Ram

Industrial Reliability and Safety Engineering
Applications and Practices
Edited by Dilbagh Panchal, Mangey Ram, Prasenjit Chatterjee, and Anish Kumar Sachdeva

Reliability and Maintenance Modeling with Optimization
Advances and Applications
Edited by Mitsutaka Kimura, Satoshi Mizutani, Mitsuhiro Imaizumi, and Kodo Ito

Intelligent Prognostics for Engineering Systems with Machine Learning Techniques
Edited by Gunjan Soni, Om Prakash Yadav, Gaurav Kumar Badhotiya, and Mangey Ram

For more information about this series, please visit: https://www.routledge.com/Advanced-Research-in-Reliability-and-System-Assurance-Engineering/book-series/CRCARRSAE

Intelligent Prognostics for Engineering Systems with Machine Learning Techniques

Edited by
Gunjan Soni
Om Prakash Yadav
Gaurav Kumar Badhotiya
Mangey Ram

CRC Press
Taylor & Francis Group
Boca Raton New York London

CRC Press is an imprint of the
Taylor & Francis Group, an **informa** business

MATLAB® is a trademark of The MathWorks, Inc. and is used with permission. The MathWorks does not warrant the accuracy of the text or exercises in this book. This book's use or discussion of MATLAB® software or related products does not constitute endorsement or sponsorship by The MathWorks of a particular pedagogical approach or particular use of the MATLAB® software.

Front cover image: Brigitte Pica2/Shutterstock

First edition published 2024
by CRC Press
2385 NW Executive Center Dr, Suite 320, Boca Raton, FL 33431

and by CRC Press
4 Park Square, Milton Park, Abingdon, Oxon, OX14 4RN

CRC Press is an imprint of Taylor & Francis Group, LLC

© 2024 selection and editorial matter Gunjan Soni, Om Prakash Yadav, Gaurav Kumar Badhotiya and Mangey Ram; individual chapters, the contributors

ISBN: 978-1-032-05436-0 (hbk)
ISBN: 978-1-032-56297-1 (pbk)
ISBN: 978-1-003-43484-9 (ebk)

DOI: 10.1201/9781003434849

Typeset in Sabon
by SPi Technologies India Pvt Ltd (Straive)

Contents

Preface

Engineering systems are critical to modern-day industrial processes, and ensuring their efficient operation is essential for maintaining productivity and safety. The field of intelligent prognostics has gained significant interest in recent years due to the potential to predict system behavior, thereby reducing downtime and maintenance costs. This book on "Intelligent Prognostics for Engineering Systems with Machine Learning" presents a comprehensive overview of the latest research in this area, with a focus on the application of machine learning techniques.

The book is comprised of 12 chapters, each of which focuses on a specific aspect of intelligent prognostics for engineering systems. The first chapter provides a bibliometric analysis of research on tool condition monitoring, highlighting the key areas of interest and the most significant contributions to this field. Subsequent chapters explore the application of machine learning in various engineering systems, including cutting tools, electrical arc furnaces, offshore topside piping, and industrial manipulators. The book covers several important topics, such as predictive maintenance, tool wear estimation, and remaining useful life prediction, along with case studies that demonstrate the practical application of machine learning techniques in real-world scenarios.

The authors of this book are experts in the field of intelligent prognostics and machine learning, and they have contributed their extensive knowledge and experience to create a valuable resource for engineers, researchers, and students alike. They provide detailed discussions of various machine learning algorithms and techniques used for prognostics, such as support vector machines, neural networks, and random forest algorithms, among others.

Overall, this book provides a comprehensive and up-to-date account of the latest research in the field of intelligent prognostics for engineering systems, which will be of great interest to researchers, engineers, and students working in this area. We hope that this book will inspire further research and innovation in this field, leading to the development of even more advanced machine learning techniques and applications for engineering systems.

About the editors

Gunjan Soni holds a BE from University of Rajasthan, MTech from IIT, Delhi, and PhD from Birla Institute of Technology and Science, Pilani, in 2012. He is presently working as an assistant professor in Department of Mechanical Engineering, Malaviya National Institute of Technology, Jaipur, Rajasthan, India. He has over 17 years of teaching experience at undergraduate and graduate levels. His areas of research interest are predictive maintenance and digital technology applications in supply chain management. He has published more than 80 papers in peer-reviewed journals including Journal of Business Research, Expert System with Applications, IEEE Transactions on Engineering Management, Production Planning and Control, Supply Chain Management: An International Journal, Annals of Operations Research, Computers and Industrial Engineering, International Journal of Logistics Research and Applications, etc. He is guest editor of special issues in journals like International Journal of Logistics Management, Sustainability, International Journal of Intelligent Enterprise, etc.

Om Prakash Yadav is a professor and Duin Endowed Fellow in the Department of Industrial and Manufacturing Engineering at North Dakota State University, Fargo. He holds a PhD in Industrial Engineering from Wayne State University, MS in Industrial Engineering from National Institute of Industrial Engineering Mumbai (India), and BS in Mechanical Engineering from Malaviya National Institute of Technology, Jaipur (India). His research interests include reliability modeling and analysis, risk assessment, design optimization, robust design, and manufacturing systems analysis. The research work of his group has been published in high-quality journals such as Reliability Engineering and Systems Safety, Journal of Risk and Reliability, Quality and Reliability Engineering International, and Engineering Management Journal. He has published over 130 papers in peer-reviewed journals and conference proceedings in the area of quality, reliability, product development, and operations management. Dr. Yadav is a recipient of the 2015 and 2018 IISE William A.J. Golomski best paper awards. He is currently a member of IISE, ASQ, SRE, and INFORMS.

Gaurav Kumar Badhotiya is currently an assistant professor in the Faculty of Management Studies, Marwadi University, Rajkot, Gujarat, India. He holds a PhD in Industrial Engineering and MTech in Manufacturing System Engineering from Malaviya National Institute of Technology, Jaipur, Rajasthan, India. His BTech is in Production and Industrial Engineering from the University College of Engineering, Kota, Rajasthan, India. His research interests are inclined toward areas in operations and supply chain management, such as supply chain resilience, production planning, circular economy, and sustainability. He has published more than 50 research articles in various peer-reviewed international journals, book chapters, and conferences proceedings. He is an editorial board member of International Journal of Mathematical, Engineering and Management Sciences. He has organized two Scopus Indexed International Conferences and a Faculty Development Program on Research Methodology and Data Analysis.

Mangey Ram holds a Ph.D. degree major in Mathematics and minor in Computer Science from G.B. Pant University of Agriculture and Technology, Pantnagar, India (2008). He is currently a research professor at Graphic Era (Deemed to be University), Dehradun, India, and a visiting professor at Peter the Great St. Petersburg Polytechnic University, Saint Petersburg, Russia. He is editor in chief of International Journal of Mathematical, Engineering and Management Sciences, Journal of Reliability and Statistical Studies, and Journal of Graphic Era University; series editor of six book series with Elsevier, CRC Press-A Taylor and Frances Group, Walter De Gruyter Publisher Germany, and River Publishers, and a guest editor and associate editor with various journals. He has published 300-plus publications (journal articles/books/book chapters/conference articles) in IEEE, Taylor & Francis, Springer Nature, Elsevier, Emerald, World Scientific, and many other national and international journals and conferences. Also, he has published more than 60 books (authored/edited) with international publishers like Elsevier, Springer Nature, CRC Press-A Taylor and Frances Group, Walter De Gruyter Publisher Germany, and River Publishers. His fields of research are reliability theory and applied mathematics. Dr. Ram is a senior member of the IEEE, senior life member of Operational Research Society of India, Society for Reliability Engineering, Quality and Operations Management in India, and Indian Society of Industrial and Applied Mathematics. He has been a member of the organizing committee of a number of international and national conferences, seminars, and workshops.

List of contributors

N. Bagalkot
Oslo Metropolitan University
Oslo, Norway

Nikhil Baliyan
Roorkee Institute of Technology
Roorkee, India

Ramnath Prabhu Bam
School of Mechanical Sciences
Indian Institute of Technology Goa
Goa, India

Tapash Kumar Das
Indian Institutes of Technology
Kharagpur, India

Rishabh Dwivedi
Malaviya National Institute of
  Technology
Jaipur, India

Rajesh S. Prabhu Gaonkar
School of Interdisciplinary Life Sciences
Indian Institute of Technology Goa
Goa, India

Clint Pazhayidam George
School of Mathematics and
  Computer Sciences
Indian Institute of Technology, Goa
Goa, India

Neeraj Kumar Goyal
Indian Institutes of Technology
Kharagpur, India

Nupur Goyal
Graphic Era Deemed to be
  University,
Dehradun, India

Rishabh Gupta
System Dynamics Lab, Department
  of Mechanical Engineering
Indian Institute of Technology
Indore, India

Pavan Kumar Kankar
System Dynamics Lab, Department
  of Mechanical Engineering
Indian Institute of Technology
Indore, India

A. Keprate
Oslo Metropolitan University
Oslo, Norway

TechnipFMC, Lysaker, Norway

A. Kharola
Graphic Era Deemed to be
  University
Dehradun, India

Devesh Kumar
Malaviya National Institute of
    Technology
Jaipur, India

P. Kumar
National Institute of Technology
    Kurukshetra
Kurukshetra, India

Rajesh Kumar
Malaviya National Institute of
    Technology
Jaipur, India

Ankur Miglani
Microfluidics and Droplet
    Dynamics Lab, Department of
    Mechanical Engineering
Indian Institute of Technology
Indore, India

Ashutosh Mishra
Malaviya National Institute of
    Technology,
Jaipur, India

Murari Lal Mittal
Malaviya National Institute of
    Technology
Jaipur, India

Namrata Mohanty
Indian Institutes of Technology
Kharagpur, India

Amitkumar Patil
Malaviya National Institute of
    Technology
Jaipur, India

Pawan
Malaviya National Institute of
    Technology
Jaipur, India

Jatin Prakash
System Dynamics Lab, Department
    of Mechanical Engineering
Indian Institute of Technology,
    Indore
Indore, India

Navin Rajpurohit
Malaviya National Institute of
    Technology
Jaipur, India

Ganpati Kumar Roy
Sharda University
Greater Noida, India

Vikas Kumar Roy
Ruprecht Karl University of
    Heidelberg
Heidelberg, Germany

Jeetesh Sharma
Malaviya National Institute of
    Technology
Jaipur, India

Bharat Singh
Malaviya National Institute of
    Technology
Jaipur, India

Gunjan Soni
Malaviya National Institute of
    Technology
Jaipur, India

Vasu Thakur
Ruprecht Karl University of
    Heidelberg
Heidelberg, Germany

J. Vaishnavi
Malaviya National Institute of
    Technology
Jaipur, India

Ankit Vijayvargiya
Swami Keshvanand Institute of
    Technology, Management &
    Gramothan
Jaipur, India

O.P. Yadav
North Dakota State University
Fargo, ND USA

Chapter 1

A bibliometric analysis of research on tool condition monitoring

Jeetesh Sharma, Murari Lal Mittal, and Gunjan Soni
Malaviya National Institute of Technology, Jaipur, India

CONTENTS

1.1 INTRODUCTION

The bibliometric study provides current trends in a particular research area and provides the overall structure of the research, which helps future research work. The term "bibliometrics" was first used by Pritchard (1969) in his article "Statistical Bibliography or Bibliometrics," published in the "Journal of Documentation." Bibliometrics as "the study and measurement of the publication pattern of all forms of written communication and their author" was described by Potter (1981). Bibliometric research is a statistical examination of publications to evaluate the impact of output. It is a quantitative method of literature review used to determine research topic trends and assess the structure of research-based publication analysis.

Condition monitoring is the procedure of observing a parameter of a condition in a machine to determine a notable change that manifests as a developing fault. TCM employs sensors and intelligence to anticipate and avoid

DOI: 10.1201/9781003434849-1

1

unfavorable machine and tool conditions (Sick 2002). Tool wear, tool break-age, etc., are adverse conditions while machining, affecting the part's dimensional accuracy and tool life. The most dominant factor in machining is the uninterrupted utilization of tools.

Two methods are used to monitor tool wear and breakage: direct (offline) and indirect (online). Radioactivity, electric resistance, optics, displacement, etc., are some direct measuring methods in which a microscope measures the variables of the tool, etc. Dimensional changes can be measured accurately by direct measuring methods. Direct measuring methods are usually carried out offline and punctuate conventional machining operations. However indirect method makes use of several sensor signals (acoustic emission, cutting force, vibrations, spindle motor power, spindle motor current, temperature, vision, etc.) to measure the tool wear (Kurada and Bradley 1997).

The TCM system flowchart is presented in Figure 1.1. The TCM comprises software and hardware part. The signal acquisition part appertains to the hardware part, and the rest of the procedure, including signal processing, feature extraction, feature selection, decision-making algorithms, and tool conditioning, belongs to the software part (Siddhpura and Paurobally 2013). The acquired signals extracted features in the time, frequency, and time-frequency domains (Shankar, Mohanraj, and Rajasekar 2019; Zhu and Wong 2009; Hsieh, Lu, and Chiou 2012). The response was illustrated as a function of time in time domain analysis, and time domain data were converted into frequency domain through a fast Fourier transform algorithm to examine the frequency components (Jemielniak and Arrazola 2008; Dimla and Lister 2000). The wavelet analysis dispenses with time-frequency domain analysis (Zhang, Yuan, and Nie 2015). Diverse strategies have been considered for automating the TCM system, including artificial neural network (ANN) (Hong, Rahman, and Zhou 1996; Shankar, Mohanraj, and Pramanik 2019), support vector machine (SVM) (Kilundu, Dehombreux, and Chiementin 2011; Wang et al. 2014; Widodo and Yang 2007; Saimurugan et al. 2011), support vector regression (SVR) (Benkedjouh et al. 2015), fuzzy logic (Li 2002; Li et al. 2004; Shankar and Mohanraj 2015), hidden Markov model (Wang and Wang 2012; Boutros and Liang 2011), decision trees (Elangovan et al. 2011; Elangovan, Ramachandran, and Sugumaran 2010), etc.

1.2 DATA COLLECTION AND RESEARCH METHODOLOGY

We have collected data from the most preferred archives: Scopus. The keyword used for the research is: "Tool condition monitoring," performed on 4 November 2020. We considered the data from 2001 to 2020. From Scopus, we recovered the author, title, citation record, h-index, etc. Scopus showed 772 documents. Of the 772 papers in Scopus major category were Articles

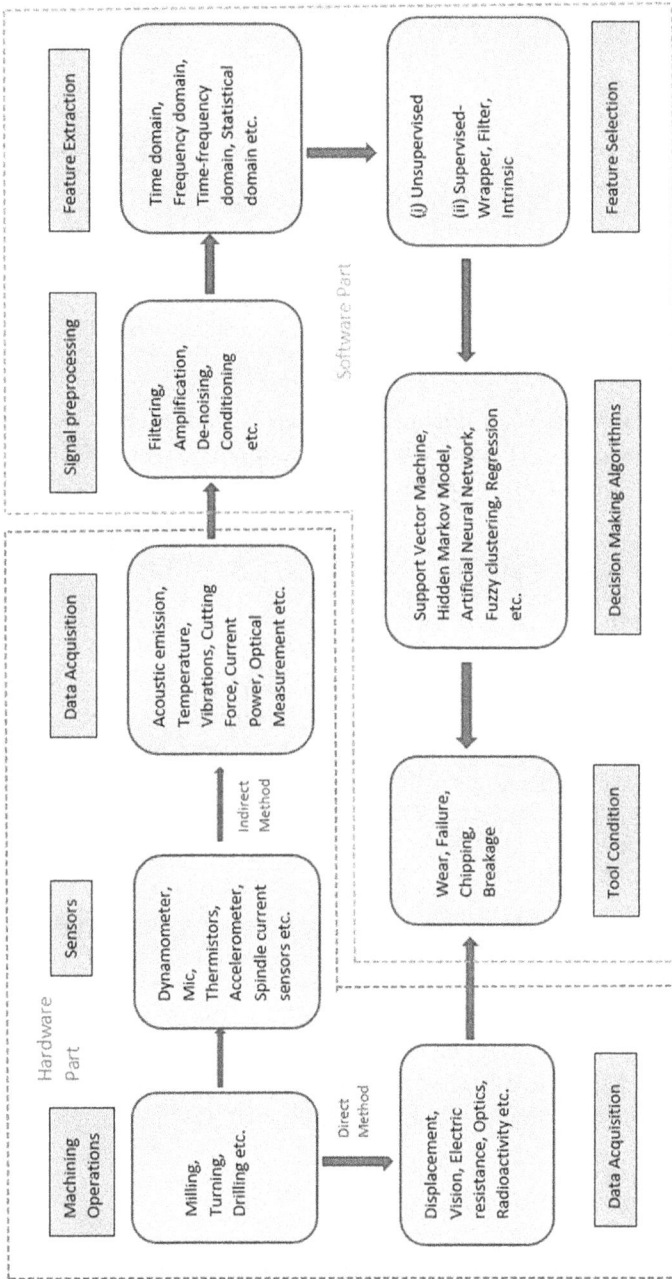

Figure 1.1 Flowchart of a tool condition monitoring (TCM) system.

Table 1.1 Types of publications in Scopus

Document types	Total numbers	% Contribution
Article	466	60.36
Conference paper	267	34.58
Review	14	1.81
Book chapter	12	1.55
Conference review	10	1.29
Article in press	1	0.12
Book	1	0.12
Erratum	1	0.12

(466), 60.36% of the total documents. Other categories were Conference paper (267), Review (14), Book chapter (2), Conference review (10), Article in press (1), Book (1), and Erratum (1). All the document types are assembled in Table 1.1.

1.3 BIBLIOMETRIC ANALYSIS

The most cited authors, the most cited manuscripts, the most cited countries, the most relevant sources, the most relevant keywords, the authors' dominance ranking, the h-index of the top ten most productive authors, and more are all covered here.

1.3.1 Research growth

Condition monitoring has been gaining expeditious attention since its establishment. Figure 1.2 shows the total number of publications in Scopus. As shown in the figure, the number of publications was less than 20 in 2001 and reached more than 70 articles in 2018. The average total citations per year have been maximum in the year 2002. The decreasing slope after 2015 indicates that the new articles have fewer citations, and their number of citations will increase in the upcoming years.

1.3.2 Most productive authors

In this section, to create the list of most productive authors, we determined the top ten authors based on the overall number of publications, as shown in Table 1.2. HONG GS is the most influential author with the highest number of articles, 24. DUTTA S is at the bottom of the list with 10 articles.

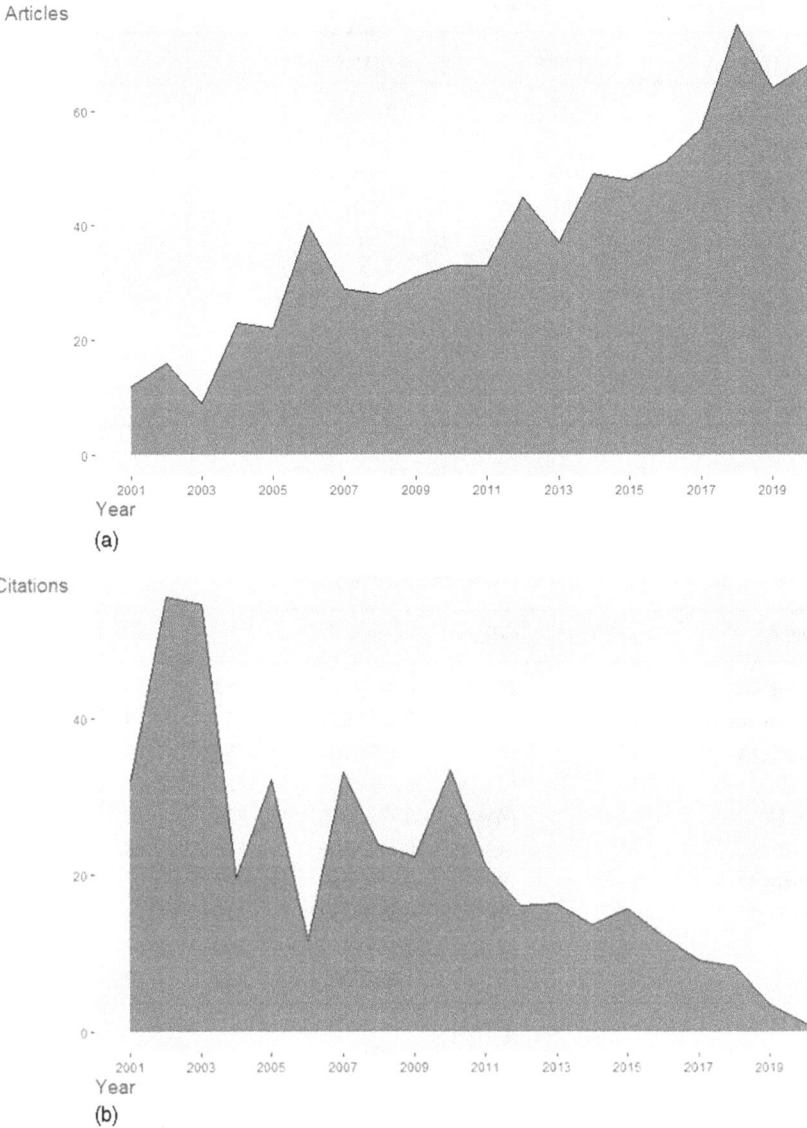

Figure 1.2 (a) The graph shows the trend of the number of publications; (b) the figure depicts average total citations per year.

1.3.2.1 h-index of the first ten most productive authors

The h-index tries to measure the citation impact and productivity of scholars' publications. As shown in Table 1.3, Hong Gs is at the top of the list, with 13 h-index value, 814 citations, and a total of 24 number of publications. Jemielniak K has an h-index value of 12 with the highest number of

Table 1.2 Most productive authors

Authors	Articles	Article authors	Articles fractionalized
Hong Gs	24	Na Na	10.00
Jemielniak K	19	Jemielniak K	8.33
Wong Ys	19	Hong Gs	6.33
Li X	17	Wong Ys	5.57
Pal Sk	14	Fu P	5.37
Fu P	13	Li X	4.58
Zhang L	13	Zhu K	3.83
Zhou Jh	13	Pal Sk	3.78
Sun J	12	Hope Ad	3.67
Dutta S	10	Heyns Ps	3.17

Table 1.3 h-Index of the first ten most productive authors

Author	h-index	g-index	m-index	TC	NP	Year
Hong Gs	13	24	0.6842105	814	24	2002
Jemielniak K	12	19	0.6000000	1275	19	2001
Wong Ys	12	19	0.6000000	794	19	2001
Li X	10	17	0.6250000	435	17	2005
Pal Sk	12	14	0.7500000	378	14	2005
Fu P	3	4	0.1764706	24	13	2004
Zhang L	4	9	0.2105263	97	12	2002
Zhou Jh	7	13	0.5833333	315	13	2009
Sun J	6	12	0.3529412	208	12	2004
Dutta S	9	10	1.0000000	332	10	2012

citations, 1,275, and a total of 19 number of publications (Egghe 2006). Here, TC represents times cited, and NP represents the number of papers. The originator of the h-index establishes the m-index. The m-index is computed by dividing the h-index by the number of years since the first publication of the researcher. The m-index normalizes the h-index value, making it easier to compare early- and late-stage researchers. Dutta S has the highest m index value. In the case of the h-index, when publications are recorded in decreasing order of citations received, the g-index is the most impressive number, recognizing a total of g^2 citations. Hong Gs has the highest value of the g-index of 24, and Fu P has the lowest value of 4.

Table 1.4 Authors' dominance ranking

Author	Dominance factor	Total articles	Single-authored	Multi-authored	First-authored	Rank by articles	Rank by DF
Fu P	0.7692308	13	0	13	10	3	1
Zhu K	0.7000000	10	0	10	7	7	2
Sun J	0.6666667	12	0	12	8	6	3
Rizal M	0.6666667	9	0	9	6	9	3
Patra K	0.6250000	9	1	8	5	9	5
Dutta S	0.6000000	10	0	10	6	7	6
Jemielniak K	0.3333333	19	4	15	5	1	7
Li X	0.2352941	17	0	17	4	2	8
Zhang L	0.1538462	13	0	13	2	3	9
Zhou Jh	0.1538462	13	0	13	2	3	9

1.3.3 Authors' dominance ranking

The dominance factor designates a fragment of multi-authored research articles in which an intellectual emerges as the first author. Fu P is at the top of the list with 0.7692308 dominance factor, as shown in Table 1.4. Sun J and Rizal M are at position 3 because of the similar value of 0.6666667 dominance factor. Zhang L and Zhou Jh are at position 9 with the same value of 0.1538462 dominance factor. This analysis, according to the dominance factor, assists us in identifying the authors' ranking more finely.

1.3.4 Top manuscripts per citations

In this particular section, the top ten manuscripts are listed in Table 1.5, according to the total number of citations from Scopus. Teti R, 2010, Cirp Ann Manuf Technol is the most cited manuscript with a total number of 744 citations and 67.64 citations per year, followed by Sick B, 2002, mech syst signal process with a total number of 333 citations and 17.53 citations per year. Salgado Dr, 2007, Int J Mach Tools Manuf is at the bottom of the list with a total number of 133 citations and 9.50 citations per year.

1.3.5 Country-wise analysis

In this section, the top ten countries are listed in Table 1.6, according to the total number of citations. China is at the top of the list with 1,183 citations and 12.59 average article citations, and Poland is at the bottom with a total number of 349 citations and 26.85 average article citations. However, Germany has the most average article citations, 54.29, with 380 total citations.

Table 1.5 Top manuscripts per citations

Paper	DOI	TC	TC per year
Teti R, 2010, Cirp Ann Manuf Technol	10.1016/j.cirp.2010.05.010	744	67.64
Sick B, 2002, Mech Syst Signal Process	10.1006/mssp.2001.1460	333	17.53
Rehorn Ag, 2005, Int J Adv Manuf Technol	10.1007/s00170-004-2038-2	269	16.81
Zhu Kp, 2009, Int J Mach Tools Manuf	10.1016/j.ijmachtools.2009.02.003	267	22.25
Jantunen E, 2002, Int J Mach Tools Manuf	10.1016/S0890-6955(02)00040-8	242	12.74
Ghosh N, 2007, Mech Syst Signal Process	10.1016/j.ymssp.2005.10.010	229	16.36
Abu-Mahfouz I, 2003, Int J Mach Tools Manuf	10.1016/S0890-6955(03)00023-3	156	8.67
Benkedjouh T, 2015, J Intell Manuf	10.1007/s10845-013-0774-6	149	24.83
Shao H, 2004, Int J Mach Tools Manuf	10.1016/j.ijmachtools.2004.05.003	141	8.29
Salgado Dr, 2007, Int J Mach Tools Manuf	10.1016/j.ijmachtools.2007.04.013 133	133	9.50

Table 1.6 Total citations per country

Country	Total citations	Average article citations
China	1,183	12.59
Singapore	1,090	33.03
India	1,087	17.25
USA	918	21.86
Canada	628	31.40
United Kingdom	486	27.00
Spain	458	38.17
Italy	419	26.19
Germany	380	54.29
Poland	349	26.85

1.3.5.1 Country-wise collaboration

The corresponding author's countries are listed below in Table 1.7. It also consists of single-country publication (SCP), the multiple-country publication (MCP), and multiple-country publication ratios (MCP Ratio). The indexes SCP and MCP represent that the particular research regulated is

Table 1.7 Corresponding author's countries

Country	Articles	Freq	SCP	MCP	MCP ratio
China	94	0.2156	79	15	0.1596
India	63	0.1445	59	4	0.0635
USA	42	0.0963	37	5	0.1190
Singapore	33	0.0757	29	4	0.1212
Canada	20	0.0459	14	6	0.3000
UK	18	0.0413	16	2	0.1111
Italy	16	0.0367	12	4	0.2500
Malaysia	14	0.0321	10	4	0.2857
Poland	13	0.0298	11	2	0.1538
Turkey	13	0.0298	12	1	0.0769

inter-country collaborated or intra-country collaborated. Figure 1.3 indicates the country-wise collaboration. China claims the highest number of articles and intra-country (SCP) and inter-country (MCP) collaboration. China is at the top of the table with 94 papers and 79, 15, 0.1596 SCP, MCP, and MCP ratios, respectively. If the MCP Ratio value is greater than zero, that means at least one article has collaborated internationally.

SCP: single-country publications (intra-country collaboration)

MCP: multiple-country publications (inter-country collaboration)

MCP Ratio: multiple-country publication ratios (inter-country collaboration index)

1.3.6 Most relevant sources

In this section, we have taken out the top ten relevant sources, which are publishing works in this area, and then sorted them on the basis of the number of articles. *International Journal of Advanced Manufacturing Technology* has the most, 88 articles, and *Measurement: Journal of the International Measurement Confederation* is in tenth position with 11 articles, in Table 1.8. *Journal of Intelligent Manufacturing, Journal of Materials Processing Technology*, and *Procedia Cirp* are at the same position with 15 articles. This analysis represents the exploration TCM has acquired through these publications over the years.

1.3.7 Keyword analysis

In this section, we listed the most relevant keywords and the corresponding number of articles on those particular keywords in Table 1.9. Here author keywords (DE) represent the authors' keywords frequency distribution, and Keywords-Plus (ID) represents the keywords frequency distribution

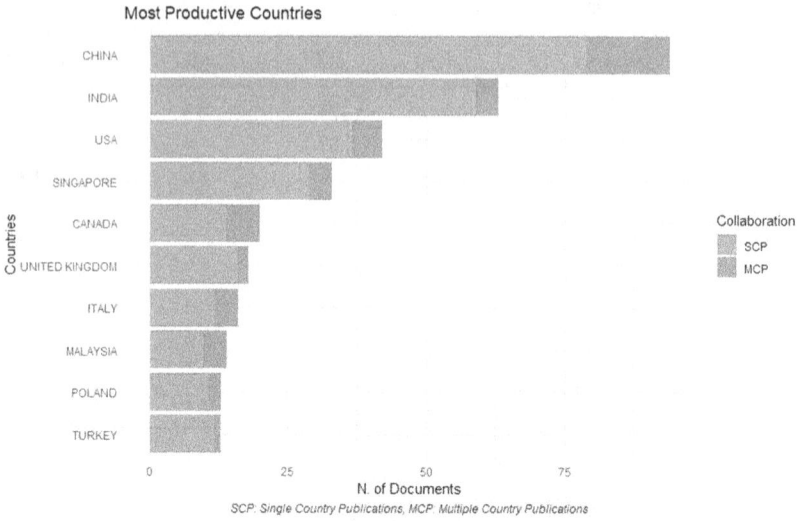

(a)

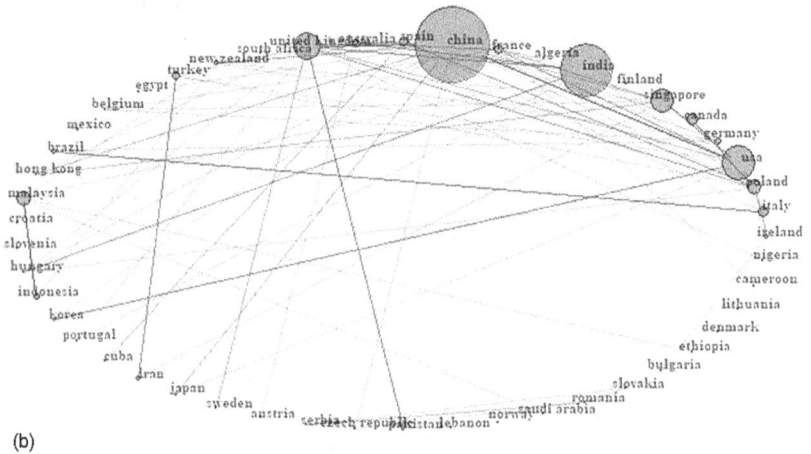

(b)

Figure 1.3 (a) Graph showing most productive countries in the given area based on MCP and SCP ratios, (b) represents the collaboration links between countries. The most productive countries such as China, India, the USA, and Singapore are the most collaborative countries, and

Country Collaboration Map

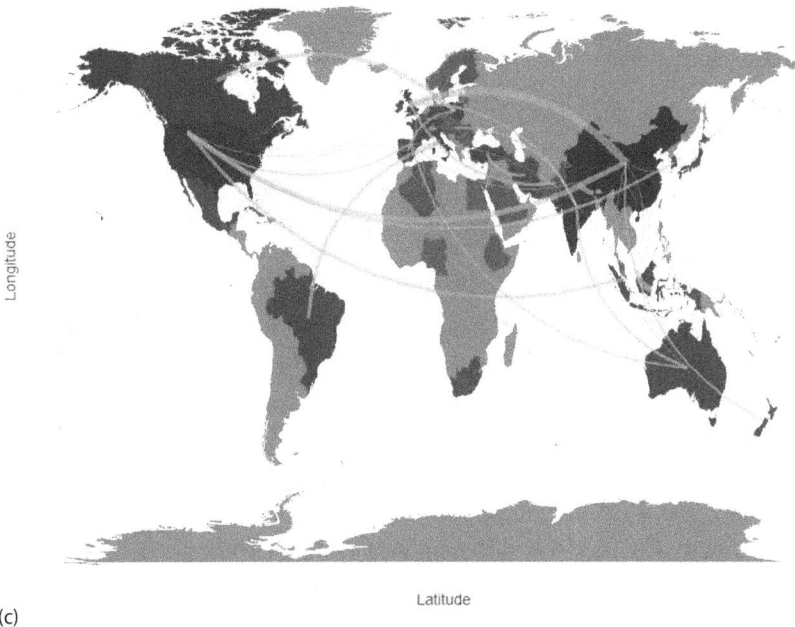

(c)

Figure 1.3 (continued) (c) countries collaborating for research in TCM are shown.

Table 1.8 Most relevant sources

Sources	Articles
International Journal of Advanced Manufacturing Technology	88
Proceedings of the Institution of Mechanical Engineers Part B: Journal of Engineering Manufacture	23
International Journal of Machine Tools and Manufacture	21
Mechanical Systems and Signal Processing	16
Journal of Intelligent Manufacturing	15
Journal of Materials Processing Technology	15
Procedia Cirp	15
Advanced Materials Research	14
Materials Today: Proceedings	12
Measurement: Journal of the International Measurement Confederation	11

Table 1.9 Most relevant keywords

Author keywords (DE)	Articles	Keywords-plus (ID)	Articles
Tool Condition Monitoring	297	Condition Monitoring	524
Tool Wear	133	Tool Condition Monitoring	486
Acoustic Emission	55	Cutting Tools	354
Condition Monitoring	37	Wear Of Materials	330
Vibration	36	Machine Tools	180
Monitoring	33	Tool Wear	168
Flank Wear	29	Milling (Machining)	167
Milling	29	Neural Networks	116
Cutting Force	26	Machining	104
Drilling	26	Cutting	102

correlated to the manuscript by SCOPUS. TCM and condition monitoring are at the top of the lists, with a frequency of 297 and 524, respectively.

The keywords used by different authors are represented in a network diagram in Figure 1.4. These keywords shown in the network diagram are connected closely with TCM.

1.3.8 Co-citation analysis

Co-citation is the recurrence with which two articles are cited together by other articles. If at least one other article cites two articles in common, they are considered to be co-cited. Figure 1.5 represents the prominent co-cited authors of TCM research. By co-citation analysis, the predominant knowledge bases of the TCM can be established proficiently and effortlessly from cited references.

1.4 CONCLUSION

A bibliometric research analysis on TCM is introduced with information associated with the most productive authors, countries, most relevant keywords, highly cited publications, etc. Scopus database was used for the data collection during this analysis. The study shows that most documents are articles, revealing accomplished publications. Moreover, HONG GS is the most productive author with the highest number of articles published and has the highest value of h-index and g-index. FU P has the highest value of the dominance factor. China claims the highest number of articles and intra-country (SCP) and inter-country (MCP) collaboration. The *International Journal of Advanced Manufacturing Technology* is the most relevant source

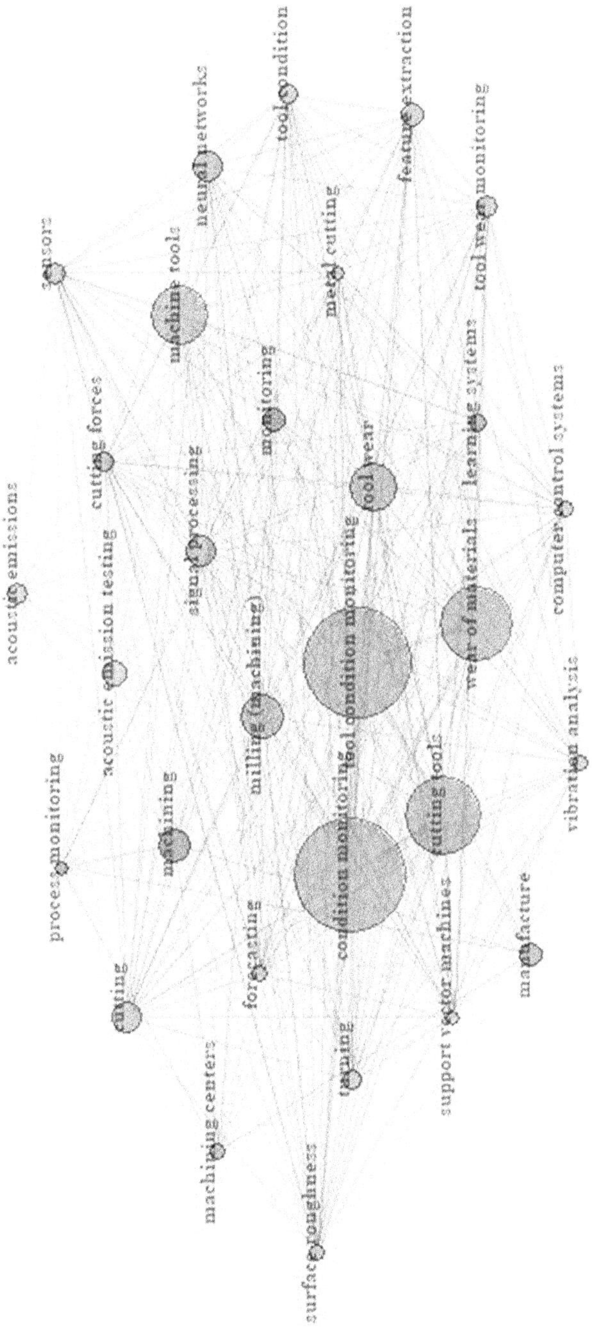

Figure 1.4 Keyword co-occurrence.

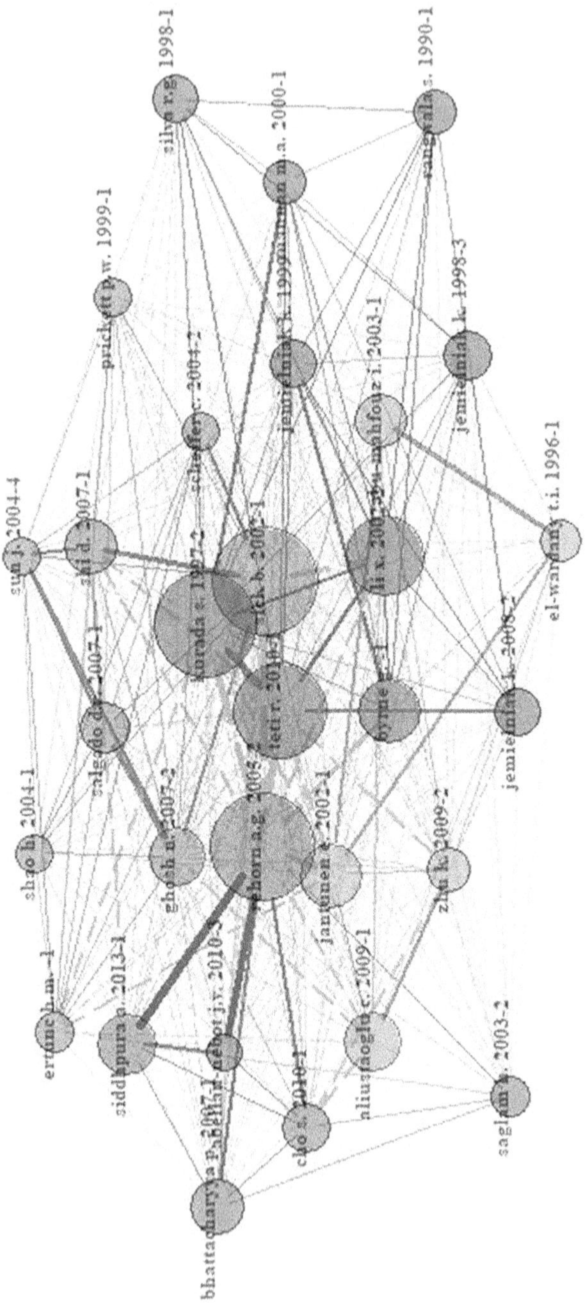

Figure 1.5 Co-citation network.

with the highest number of articles. This analysis provides an intrinsic structure of publications on TCM. Visualization diagrams of keywords co-occurrence and co-citation network were also depicted.

As with most studies, this analysis offers fascinating insights but is also pretentious by some limitations. Google Scholar results were not taken into consideration because the free ingress to scholars and publications is also in different languages other than English, and some unpredictability is also there. Only the Scopus database is considered for the data collection, but there are also other popular databases. Hence, for future purposes, more analysis with other databases could be done. The discernment obtained from this research analysis has implications for academic scholars. For establishing new future work, this study will provide a better direction.

APPENDIX I

This section provides the top 30 highly cited papers in Scopus. Table A.1 contains the author's name, year of publication, title, source, and the total number of citations (TC). Teti et al. 2010 got the most citations with a total of 747. It is followed by Sick 2002, with 333 citations. Kuljanic and Sortino 2005 are at the bottom of the list with a total of 71 citations. This table helps to provide the most influential research in the area of TCM.

Table A.1 Top 30 influential papers

Authors and Year	Title	Source	TC
(Teti et al. 2010)	"Advanced monitoring of machining operations"	*CIRP Annals – Manufacturing Technology*	747
(Sick 2002)	"On-line and indirect tool wear monitoring in turning with artificial neural networks: A review of more than a decade of research"	*Mechanical Systems and Signal Processing*	333
(Rehorn, Jiang, and Orban 2005)	"State-of-the-art methods and results in tool condition monitoring: A review"	*International Journal of Advanced Manufacturing Technology*	271
(Zhu, Wong, and Hong 2009a)	"Wavelet analysis of sensor signals for tool condition monitoring: A review and some new results"	*International Journal of Machine Tools and Manufacture*	270

(Continued)

Table A.1 (Continued)

Authors and Year	Title	Source	TC
(Jantunen 2002)	"A summary of methods applied to tool condition monitoring in drilling"	*International Journal of Machine Tools and Manufacture*	242
(Ghosh et al. 2007)	"Estimation of tool wear during CNC milling using neural network-based sensor fusion"	*Mechanical Systems and Signal Processing*	233
(Abu-Mahfouz 2003)	"Drilling wear detection and classification using vibration signals and artificial neural network"	*International Journal of Machine Tools and Manufacture*	156
(Benkedjouh et al. 2015)	"Health assessment and life prediction of cutting tools based on support vector regression"	*Journal of Intelligent Manufacturing*	150
(Shao, Wang, and Zhao 2004)	"A cutting power model for tool wear monitoring in milling"	*International Journal of Machine Tools and Manufacture*	141
(Salgado and Alonso 2007)	"An approach based on current and sound signals for in-process tool wear monitoring"	*International Journal of Machine Tools and Manufacture*	133
(Siddhpura and Paurobally 2013)	"A review of flank wear prediction methods for tool condition monitoring in a turning process"	*International Journal of Advanced Manufacturing Technology*	121
(Nouri et al. 2015)	"Real-time tool wear monitoring in milling using a cutting condition independent method"	*International Journal of Machine Tools and Manufacture*	119
(Zhu, Wong, and Hong 2009b)	"Multi-category micro-milling tool wear monitoring with continuous hidden Markov models"	*Mechanical Systems and Signal Processing*	119
(Huang and Liang 2003)	"Cutting forces modeling considering the effect of tool thermal property—Application to CBN hard turning"	*International Journal of Machine Tools and Manufacture*	116

(Continued)

Table A.1 (Continued)

Authors and Year	Title	Source	TC
(Alonso and Salgado 2008)	"Analysis of the structure of vibration signals for tool wear detection"	*Mechanical Systems and Signal Processing*	112
(Bhattacharyya, Sengupta, and Mukhopadhyay 2007)	"Cutting force-based real-time estimation of tool wear in face milling using a combination of signal processing techniques"	*Mechanical Systems and Signal Processing*	112
(Jemielniak and Arrazola 2008)	"Application of AE and cutting force signals in tool condition monitoring in micro-milling"	*CIRP Journal of Manufacturing Science and Technology*	102
(Dimla Snr. 2002)	"The correlation of vibration signal features to cutting tool wear in a metal turning operation"	*International Journal of Advanced Manufacturing Technology*	98
(Dutta et al. 2013)	"Application of digital image processing in tool condition monitoring: A review"	*CIRP Journal of Manufacturing Science and Technology*	96
(Marinescu and Axinte 2008)	"A critical analysis of the effectiveness of acoustic emission signals to detect tool and workpiece malfunctions in milling operations"	*International Journal of Machine Tools and Manufacture*	95
(Balazinski et al. 2002)	"Tool condition monitoring using artificial intelligence methods"	*Engineering Applications of Artificial Intelligence*	94
(Scheffer and Heyns 2001)	"Wear monitoring in turning operations using vibration and strain measurements"	*Mechanical Systems and Signal Processing*	94
(Sun et al. 2004)	"Multi classification of tool wear with support vector machine by manufacturing loss consideration"	*International Journal of Machine Tools and Manufacture*	92

(*Continued*)

Table A.1 (Continued)

Authors and Year	Title	Source	TC
(Zhou et al. 2011)	"Tool wear monitoring using acoustic emissions by dominant-feature identification"	*IEEE Transactions on Instrumentation and Measurement*	88
(Huang and Liang 2005)	"Modeling of cutting forces under hard turning conditions considering tool wear effect"	*Journal of Manufacturing Science and Engineering, Transactions of the ASME*	87
(Peng and Xu 2014)	"Energy-efficient machining systems: A critical review"	*International Journal of Advanced Manufacturing Technology*	84
(Chen and Li 2007)	"Acoustic emission method for tool condition monitoring based on wavelet analysis"	*International Journal of Advanced Manufacturing Technology*	78
(Li et al. 2009)	"Fuzzy neural network modeling for tool wear estimation in dry milling operation"	"Annual Conference of the Prognostics and Health Management Society"	72
(Kuljanic and Sortino 2005)	"TWEM, a method based on cutting forces - Monitoring tool wear in face milling"	*International Journal of Machine Tools and Manufacture*	71

REFERENCES

Abu-Mahfouz, Issam. 2003. "Drilling Wear Detection and Classification Using Vibration Signals and Artificial Neural Network." *International Journal of Machine Tools and Manufacture* 43 (7): 707–20. https://doi.org/10.1016/S0890-6955(03)00023-3.

Alonso, F.J., and D.R. Salgado. 2008. "Analysis of the Structure of Vibration Signals for Tool Wear Detection." *Mechanical Systems and Signal Processing* 22 (3): 735–48. https://doi.org/10.1016/j.ymssp.2007.09.012.

Balazinski, Marek, Ernest Czogala, Krzysztof Jemielniak, and Jacek Leski. 2002. "Tool Condition Monitoring Using Artificial Intelligence Methods." *Engineering Applications of Artificial Intelligence* 15 (1): 73–80. https://doi.org/10.1016/S0952-1976(02)00004-0.

Benkedjouh, T., K. Medjaher, N. Zerhouni, and S. Rechak. 2015. "Health Assessment and Life Prediction of Cutting Tools Based on Support Vector Regression." *Journal of Intelligent Manufacturing* 26 (2): 213–23. https://doi.org/10.1007/s10845-013-0774-6.

Bhattacharyya, P., D. Sengupta, and S. Mukhopadhyay. 2007. "Cutting Force-Based Real-Time Estimation of Tool Wear in Face Milling Using a Combination of Signal Processing Techniques." *Mechanical Systems and Signal Processing* 21 (6): 2665–83. https://doi.org/10.1016/j.ymssp.2007.01.004.

Boutros, Tony, and Ming Liang. 2011. "Detection and Diagnosis of Bearing and Cutting Tool Faults Using Hidden Markov Models." *Mechanical Systems and Signal Processing* 25 (6): 2102–24. https://doi.org/10.1016/j.ymssp.2011.01.013.

Chen, Xiaozhi, and Beizhi Li. 2007. "Acoustic Emission Method for Tool Condition Monitoring Based on Wavelet Analysis." *International Journal of Advanced Manufacturing Technology* 33 (9–10): 968–76. https://doi.org/10.1007/s00170-006-0523-5.

Dimla, D.E., and P.M. Lister. 2000. "On-Line Metal Cutting Tool Condition Monitoring. I: Force and Vibration Analyses." *International Journal of Machine Tools and Manufacture* 40 (5): 739–68. https://doi.org/10.1016/S0890-6955(99)00084-X.

Dimla, Snr, D.E. 2002. "The Correlation of Vibration Signal Features to Cutting Tool Wear in a Metal Turning Operation." *International Journal of Advanced Manufacturing Technology* 19 (10): 705–13. https://doi.org/10.1007/s001700 200080.

Dutta, S., S.K. Pal, S. Mukhopadhyay, and R. Sen. 2013. "Application of Digital Image Processing in Tool Condition Monitoring: A Review." *CIRP Journal of Manufacturing Science and Technology* 6 (3): 212–32. https://doi.org/10.1016/j.cirpj.2013.02.005.

Egghe, Leo. 2006. "Theory and Practise of the G-Index." *Scientometrics* 69 (1): 131–52. https://doi.org/10.1007/s11192-006-0144-7.

Elangovan, M., S. Babu Devasenapati, N.R. Sakthivel, and K.I. Ramachandran. 2011. "Evaluation of Expert System for Condition Monitoring of a Single Point Cutting Tool Using Principle Component Analysis and Decision Tree Algorithm." *Expert Systems with Applications* 38 (4): 4450–59. https://doi.org/10.1016/j.eswa.2010.09.116.

Elangovan, M., K.I. Ramachandran, and V. Sugumaran. 2010. "Studies on Bayes Classifier for Condition Monitoring of Single Point Carbide Tipped Tool Based on Statistical and Histogram Features." *Expert Systems with Applications* 37 (3): 2059–65. https://doi.org/10.1016/j.eswa.2009.06.103.

Ghosh, N., Y.B. Ravi, A. Patra, S. Mukhopadhyay, S. Paul, A.R. Mohanty, and A.B. Chattopadhyay. 2007. "Estimation of Tool Wear during CNC Milling Using Neural Network-Based Sensor Fusion." *Mechanical Systems and Signal Processing*, 21 (1): 466–79.

Hong, G.S., M. Rahman, and Q. Zhou. 1996. "Using Neural Network for Tool Condition Monitoring Based on Wavelet Decomposition." *International Journal of Machine Tools and Manufacture* 36 (5): 551–66. https://doi.org/10.1016/0890-6955(95)00067-4.

Hsieh, Wan Hao, Ming Chyuan Lu, and Shean Juinn Chiou. 2012. "Application of Backpropagation Neural Network for Spindle Vibration-Based Tool Wear Monitoring in Micro-Milling." *International Journal of Advanced Manufacturing Technology* 61 (1–4): 53–61. https://doi.org/10.1007/s00170-011-3703-x.

Huang, Y, and S. Y. Liang. 2005. "Modeling of Cutting Forces under Hard Turning Conditions Considering Tool Wear Effect." *The Journal of Manufacturing Science and Engineering* 127 (2): 262–70.

Huang, Y., and Steven Y. Liang. 2003. "Cutting Forces Modeling Considering the Effect of Tool Thermal Property—Application to CBN Hard Turning." *International Journal of Machine Tools and Manufacture* 43 (3): 307–15. https://doi.org/10.1016/S0890-6955(02)00185-2.

Jantunen, Erkki. 2002. "A Summary of Methods Applied to Tool Condition Monitoring in Drilling." *International Journal of Machine Tools and Manufacture* 42 (9): 997–1010. https://doi.org/10.1016/S0890-6955(02)00040-8.

Jemielniak, K., and P.J. Arrazola. 2008. "Application of AE and Cutting Force Signals in Tool Condition Monitoring in Micro-Milling." *CIRP Journal of Manufacturing Science and Technology* 1 (2): 97–102. https://doi.org/10.1016/j.cirpj.2008.09.007.

Kilundu, Bovic, Pierre Dehombreux, and Xavier Chiementin. 2011. "Tool Wear Monitoring by Machine Learning Techniques and Singular Spectrum Analysis." *Mechanical Systems and Signal Processing* 25 (1): 400–415. https://doi.org/10.1016/j.ymssp.2010.07.014.

Kuljanic, E., and M. Sortino. 2005. "TWEM." *A Method Based on Cutting Forces---Monitoring Tool Wear in Face Milling* 45 (1): 29–34.

Kurada, Satya, and Colin Bradley. 1997. "A Review of Machine Vision Sensors for Tool Condition Monitoring." *Computers in Industry* 34 (1): 55–72. https://doi.org/10.1016/s0166-3615(96)00075-9.

Li, X. 2002. "A Brief Review: Acoustic Emission Method for Tool Wear Monitoring during Turning." *International Journal of Machine Tools and Manufacture* 42 (2): 157–65. https://doi.org/10.1016/S0890-6955(01)00108-0.

Li, X., H.X. Li, X.P. Guan, and R. Du. 2004. "Fuzzy Estimation of Feed-Cutting Force from Current Measurement-a Case Study on Intelligent Tool Wear Condition Monitoring. IEEE Transactions on Systems, Man, and Cybernetics." *Part C (Applications and Reviews)*, 34 (4): 506–12.

Li, X., B.S. Lim, J.H. Zhou, S. Huang, S.J. Phua, K.C. Shaw, and M.J. Er. 2009. "Fuzzy Neural Network Modelling for Tool Wear Estimation in Dry Milling Operation." *In Annual Conference of the PHM Society* (Vol. 1: 1).

Marinescu, Iulian, and Dragos A. Axinte. 2008. "A Critical Analysis of Effectiveness of Acoustic Emission Signals to Detect Tool and Workpiece Malfunctions in Milling Operations." *International Journal of Machine Tools and Manufacture* 48 (10): 1148–60. https://doi.org/10.1016/j.ijmachtools.2008.01.011.

Nouri, Mehdi, Barry K. Fussell, Beth L. Ziniti, and Ernst Linder. 2015. "Real-Time Tool Wear Monitoring in Milling Using a Cutting Condition Independent Method." *International Journal of Machine Tools and Manufacture* 89: 1–13. https://doi.org/10.1016/j.ijmachtools.2014.10.011.

Peng, T., and X. Xu. 2014. "Energy-Efficient Machining Systems: A Critical Review." *The International Journal of Advanced Manufacturing Technology* 72 (9–12): 1389–1406.

Potter, W.G. 1981. Introduction to bibliometrics. *Library Trends* 30(5).

Pritchard, A. 1969. Statistical bibliography or bibliometrics. *Journal of documentation*, 25, 348.

Rehorn, Adam G., Jin Jiang, and Peter E. Orban. 2005. "State-of-the-Art Methods and Results in Tool Condition Monitoring: A Review." *International Journal of Advanced Manufacturing Technology* 26 (7–8): 693–710. https://doi.org/10.1007/s00170-004-2038-2.

Saimurugan, M., K.I. Ramachandran, V. Sugumaran, and N.R. Sakthivel. 2011. "Multi Component Fault Diagnosis of Rotational Mechanical System Based on Decision Tree and Support Vector Machine." *Expert Systems with Applications* 38 (4): 3819–26. https://doi.org/10.1016/j.eswa.2010.09.042.

Salgado, D.R., and F.J. Alonso. 2007. "An Approach Based on Current and Sound Signals for In-Process Tool Wear Monitoring." *International Journal of Machine Tools and Manufacture* 47 (14): 2140–52. https://doi.org/10.1016/j.ijmachtools.2007.04.013.

Scheffer, C., and P.S. Heyns. 2001. "Wear Monitoring in Turning Operations Using Vibration and Strain Measurements." *Mechanical Systems and Signal Processing* 15 (6): 1185–1202. https://doi.org/10.1006/mssp.2000.1364.

Shankar, S, and T. Mohanraj. 2015. "November. Tool Condition Monitoring in Milling Using Sensor Fusion Technique." *In Proceedings of Malaysian International Tribology Conference* (Vol. 2015: 322–23).

Shankar, S., T. Mohanraj, and A. Pramanik. 2019. "Tool Condition Monitoring While Using Vegetable Based Cutting Fluids during Milling of Inconel 625." *Journal of Advanced Manufacturing Systems* 18 (4): 563–81. https://doi.org/10.1142/S0219686719500306.

Shankar, S., T. Mohanraj, and R. Rajasekar. 2019. "Prediction of Cutting Tool Wear during Milling Process Using Artificial Intelligence Techniques." *International Journal of Computer Integrated Manufacturing* 32 (2): 174–82. https://doi.org/10.1080/0951192X.2018.1550681.

Shao, H., H.L. Wang, and X.M. Zhao. 2004. "A Cutting Power Model for Tool Wear Monitoring in Milling." *International Journal of Machine Tools and Manufacture* 44 (14): 1503–9. https://doi.org/10.1016/j.ijmachtools.2004.05.003.

Sick, B. 2002. "On-Line and Indirect Tool Wear Monitoring in Turning with Artificial Neural Networks: A Review of More than a Decade of Research." *Mechanical Systems and Signal Processing* 16 (4): 487–546.

Siddhpura, A., and R. Paurobally. 2013. "A Review of Flank Wear Prediction Methods for Tool Condition Monitoring in a Turning Process." *International Journal of Advanced Manufacturing Technology* 65 (1–4): 371–93. https://doi.org/10.1007/s00170-012-4177-1.

Sun, J., M. Rahman, Y.S. Wong, and G.S. Hong. 2004. "Multiclassification of Tool Wear with Support Vector Machine by Manufacturing Loss Consideration." *International Journal of Machine Tools and Manufacture* 44 (11): 1179–87. https://doi.org/10.1016/j.ijmachtools.2004.04.003.

Teti, R., K. Jemielniak, G. O'Donnell, and D. Dornfeld. 2010. "Advanced Monitoring of Machining Operations." *CIRP Annals - Manufacturing Technology* 59 (2): 717–39. https://doi.org/10.1016/j.cirp.2010.05.010.

Wang, G.F., Y.W. Yang, Y.C. Zhang, and Q.L. Xie. 2014. "Vibration Sensor Based Tool Condition Monitoring Using ν Support Vector Machine and Locality Preserving Projection." *Sensors and Actuators, A: Physical* 209: 24–32. https://doi.org/10.1016/j.sna.2014.01.004.

Wang, Mei, and Jie Wang. 2012. "CHMM for Tool Condition Monitoring and Remaining Useful Life Prediction." *International Journal of Advanced Manufacturing Technology* 59 (5–8): 463–71. https://doi.org/10.1007/s00170-011-3536-7.

Widodo, Achmad, and Bo Suk Yang. 2007. "Support Vector Machine in Machine Condition Monitoring and Fault Diagnosis." *Mechanical Systems and Signal Processing* 21 (6): 2560–74. https://doi.org/10.1016/j.ymssp.2006.12.007.

Zhang, Kai feng, Hui Qun Yuan, and Peng Nie. 2015. "A Method for Tool Condition Monitoring Based on Sensor Fusion." *Journal of Intelligent Manufacturing* 26 (5): 1011–26. https://doi.org/10.1007/s10845-015-1112-y.

Zhou, Jun Hong, Chee Khiang Pang, Zhao Wei Zhong, and Frank L. Lewis. 2011. "Tool Wear Monitoring Using Acoustic Emissions by Dominant-Feature Identification." *IEEE Transactions on Instrumentation and Measurement* 60 (2): 547–59. https://doi.org/10.1109/TIM.2010.2050974.

Zhu, K., and San Wong. 2009. "Y. and Hong, G." *S Wavelet Analysis of Sensor Signals for Tool Condition Monitoring: A Review and Some New Results* 49 (7–8): 537–53.

Zhu, Kunpeng P., Yoke San Wong, and Geok Soon Hong. 2009a. "Wavelet Analysis of Sensor Signals for Tool Condition Monitoring: A Review and Some New Results." *International Journal of Machine Tools and Manufacture* 49 (7–8): 537–53. https://doi.org/10.1016/j.ijmachtools.2009.02.003.

Zhu, Kunpeng, Yoke San Wong, and Geok Soon Hong. 2009b. "Multi-Category Micro-Milling Tool Wear Monitoring with Continuous Hidden Markov Models." *Mechanical Systems and Signal Processing* 23 (2): 547–60. https://doi.org/10.1016/j.ymssp.2008.04.010.

Chapter 2

Predicting restoration factor for different maintenance types

Neeraj Kumar Goyal, Tapash Kumar Das, and Namrata Mohanty

Indian Institute of Technology, Kharagpur, India

CONTENTS

2.1 INTRODUCTION

The railway industry is considered an environmental-friendly transportation mode, and its demand has been increasing over years. There is a requirement to maintain a high level of reliability, safety, availability and maintainability within the rail system. Rolling stock maintenance can be categorized as corrective maintenance (CM) and preventive maintenance (PM). In PM, a system can be restored to or retained in its good, healthy condition by the execution of a periodical inspection, detection and prevention of budding failures before the occurrence of failure. In case of CM, repair activities are carried out after the occurrence of failure (MEYKAR 1967).

Reliability analysis of a repairable system (RS) is generally carried out by assuming the repair process as either as good as new (AGAN) or as bad as old (ABAO) (Doyen and Gaudoin 2004). These are two extreme repair processes in which restoration of the system is either 0 or 1. The value for minimal repair (ABAO) is 0, and for renewal (AGAN) it is 1. These are two extreme cases of real condition. The real condition of the RS is in between these two extreme cases. This type of maintenance is mentioned as imperfect maintenance (Pham and Wang 1996; Nasr, Gasmi, and Sayadi 2013).

DOI: 10.1201/9781003434849-2

Several researches have been extensively carried out to model this repair process of RS. The most commonly used model with the idea of "virtual age" is generalized renewal process (GRP) developed by Kijima. Kijima has given two models, K-I and K-II (Nakagawa 1979; Nakagawa 1979).

The virtual age model takes into account the impact of maintenance actions by reducing the system's real age to virtual age (Trust 1989). It depends upon the amount of restoration achieved through maintenance activities performed. This amount of restoration is presented through a parameter called "restoration factor (RF)" (Malik and Khan 1979). Determination of RF simplifies the task of reliability estimation as it helps to understand impact on age reduction of system when certain part replacements are carried out. RF prediction makes estimation of failure distribution parameters practical and efficient.

GRP model parameters are generally estimated using maximum likelihood estimation (MLE) approach (Trust 1989). GRP model parameters are shape, scale and RFs of the repair actions (Yaez, Joglar, and Modarres 2002).

Most of the papers in literature have considered only one RF for PM activities and one for CM activities. These RFs are estimated on the basis of failure data, and they do not consider the repair actions carried out during maintenance (Yu, Song, and Cassady 2008; Said and Taghipour 2016). In practice, there are more than one type of PM or CM. Different maintenance types have different improvements in health of the system. Solving MLE equation for more than two RF parameters becomes challenging as closed form solutions become difficult to obtain. To overcome this problem, a method is proposed to predict RF parameters for various maintenance actions considering the maintenance actions performed for different PM and CM.

A case study of an E-traction Loco's motor is used to demonstrate the proposed model. Complex systems, such as railway systems, can be broken down into sub-system, sub-sub-system and so on to a component level. At a component level, it is easy to find RF. The analytical hierarchical process (AHP) is a multi-criteria decision-making tool that allows decision-makers to model a complex problem in a hierarchy (Triantaphyllou et al. 1997). As a result, this methodology is proposed to determine how much each component repair contributes to system health improvement.

2.2 PROPOSED MODEL

In this proposed methodology, the main concern is to prepare a model which accounts for improvement in the system health due to different types of maintenance activities. The main focus is on calculating RF for different types of maintenance, namely term overhaul (TOH), immediate overhaul (IOH) and periodic overhaul (POH).

For RSs, reliability estimation is generally carried out using virtual age models. Virtual age models consider the effect of maintenance actions by reducing system age using RF. Generally, RF is estimated from system failure data using various statistical methods. However, railway systems experience various types of maintenance actions at different times during their life cycle. To consider all these different types of actions, we need multiple RF parameters in the virtual age model. As failure data is limited, estimation of so many parameters becomes a complex problem. These RFs are representative of the effects of maintenance activities on the system. Therefore, these can be predicted from the information about the maintenance actions performed on the system.

MLE for the RF is in use, but there are many problems in calculating RF by MLE, as the MLE equations become very complex. RF from MLE assumes all types of PM as same and does not consider difference in maintenance activities. Different types of maintenance activities performed in different types of maintenance should give different restoration levels.

2.2.1 Steps of RF prediction

The proposed model considers the maintenance actions performed during a particular maintenance activity on components of the system, to predict overall RF of the system from the contribution of components replacement in system restoration. This prediction is carried out considering expert opinion through proposed AHP model.

The proposed methodology is summarized as follows:

Step 1: To study the given system and tabulate the components of the system
Step 2: To tabulate replaced components for different maintenance types
Step 3: To apply AHP for calculating weightage of each component
Step 4: To calculate RF for different types of PM and CM
Step 5: To calculate failure distribution parameters, i.e., shape parameter, scale parameter, etc.

2.2.1.1 To study the given system and tabulate the components of the system

Components of the system are studied and tabulated as shown in Table 2.1. There are four components of the system, C1–C4, as listed in the table.

2.2.1.2 To tabulate replaced components for different maintenance types

In this step, the maintenance activities performed during different types of maintenance are written down. The components, which are replaced for different types of maintenance, are tabulated as shown in Table 2.2. Here, Y represents the component replacement and N represents non-replacement

Table 2.1 Components of the system

S. No.	Components
I	CI
2	C2
3	C3
4	C4

Table 2.2 Replacement of components at different types of maintenance

Components	PMI	PM2	PM3
CI	Y	Y	Y
C2	N	Y	Y
C3	N	Y	Y
C4	N	N	Y

for the corresponding PM. In this system, three different PM activities are performed involving replacement of different components.

2.2.1.3 To apply AHP for calculating weightage of each component

In this step, first, hierarchical structure of AHP is prepared. At the top level of the hierarchy, the goal of the AHP is kept. In this case, goal is "Weightage of each component on system RF." At the intermediate level, criteria are placed, on the basis of which weightage of the components is obtained. In this case, there are four criteria: frequency of failure (FoF), functionality, safety and efficiency. At the bottom level of the hierarchy, alternatives are placed. At the end of this step, the RF for each component as weight is evaluated. In this case, alternatives are components of the system. All the three levels of AHP are illustrated in Figure 2.1.

2.2.1.4 To calculate RF for different PM and CM

To evaluate RF for different PM and CM, the first step is to prepare maintenance activity table from maintenance manual of the system shown in Table 2.2. If the component is replaced, it is represented as 'Y,' otherwise 'N'. Each maintenance type is represented by a column matrix (PM_j), with 1 representing Y and 0 representing N. These column matrices are shown in Equation (2.1). In general, for any j^{th} maintenance type, PM_j is given, as shown in Equation (2.2).

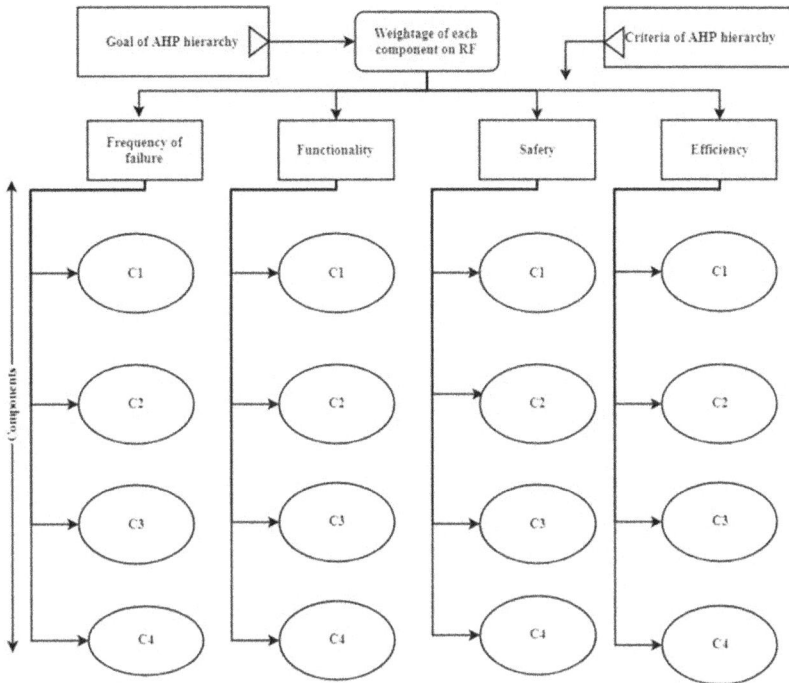

Figure 2.1 Three levels of AHP.

System RF is simply a summation of RF of components replaced in the maintenance procedure. Product matrix *P* is given in Equations (2.3)–(2.5). RF for j^{th} maintenance type PM_j is represented by RF_{PMj} and is given in Equation (2.6).

Similarly, RF for each CM is evaluated, as given in Equations (2.7)–(2.12). However, prediction of RF is not possible here as we cannot predict which component will fail. Similarly, RF for CM is prepared as shown in Table 2.3.

Table 2.3 Replacement of components at different CM instances

Components	CM_1	CM_2
C_1	Y	N
C_2	N	Y
C_3	N	N
C_4	Y	N

$$PM_1 = \begin{bmatrix} 1 \\ 0 \\ 0 \\ 0 \end{bmatrix}, PM_2 = \begin{bmatrix} 1 \\ 1 \\ 0 \\ 0 \end{bmatrix}, PM_1 = \begin{bmatrix} 1 \\ 0 \\ 0 \\ 0 \end{bmatrix} \tag{2.1}$$

$$PM_j = \begin{bmatrix} y_1 \\ y_2 \\ y_3 \\ y_4 \end{bmatrix}, \tag{2.2}$$

$$y_k = \begin{cases} 1 & \text{if component } i \text{ is replaced and} \\ 0 & \text{if component } i \text{ is not replaced,} \\ & \quad \text{for } k = 1,2,3,4 \end{cases}$$

$$P = W.PMj \tag{2.3}$$

$$P_{PMj} = \begin{bmatrix} w_1 \\ w_2 \\ w_3 \\ w_4 \end{bmatrix} \cdot \begin{bmatrix} y_1 \\ y_2 \\ y_3 \\ y_4 \end{bmatrix} = \begin{bmatrix} w_1 y_1 \\ w_2 y_2 \\ w_3 y_3 \\ w_4 y_4 \end{bmatrix} \tag{2.4}$$

$$P_{PMj} = \begin{bmatrix} p_1 \\ p_2 \\ p_3 \\ p_4 \end{bmatrix} \text{ where, } p_i = w_i \times y_i \tag{2.5}$$

$$RF_{PMj} = SUM\left[P_{PMj}\right] = SUM\left\{W.P_{PMj}\right\} = \sum_{\substack{i=1 \\ i \in PMj}}^{4 \text{ (in this case)}} \left(w_i \times y_i\right) = \sum_{\substack{i=1 \\ i \in PMj}}^{4} p_i \tag{2.6}$$

$$CM_1 = \begin{bmatrix} 1 \\ 0 \\ 0 \\ 1 \end{bmatrix}, CM_2 = \begin{bmatrix} 0 \\ 1 \\ 0 \\ 0 \end{bmatrix} \tag{2.7}$$

$$CM_k = \begin{bmatrix} y_1 \\ y_2 \\ y_3 \\ y_4 \end{bmatrix} \tag{2.8}$$

$$y_i = \begin{cases} 1 & \text{if component } i \text{ is replaced and} \\ 0 & \text{if component } i \text{ is not replaced,} \\ & \quad \text{for } k = 1,2,3,4 \end{cases}$$

It implies that for CM_1, the value of y_1 and y_4 is 1 and for rest are 0:

$$P = W.CM_k \tag{2.9}$$

$$P_{CM_k} = \begin{bmatrix} w_1 \\ w_2 \\ w_3 \\ w_4 \end{bmatrix} \cdot \begin{bmatrix} y_1 \\ y_2 \\ y_3 \\ y_4 \end{bmatrix} = \begin{bmatrix} w_1 y_1 \\ w_2 y_2 \\ w_3 y_3 \\ w_4 y_4 \end{bmatrix} \tag{2.10}$$

$$P_{CM_k} = \begin{bmatrix} p_1 \\ p_2 \\ p_3 \\ p_4 \end{bmatrix} \text{where,} p_i = w_i \times y_i, \tag{2.11}$$

$$RF_{CMi} = SUM\left[P_{PM_i}\right] = SUM\{W.CM_k\} = \sum_{\substack{i=1 \\ i \in CMi}}^{4} p_i \tag{2.12}$$

2.2.1.5 To calculate failure distribution parameters

At this step, after RF calculation for different types of PM and CM, failure distribution parameters of the system are to be determined. According to the proposed model, first we determine the virtual age at all the points of failure (CM) and PM by applying the values of RF for different types of PM and CM evaluated. Equation (2.13) satisfies the values of virtual age for corresponding PM and CM obtained from virtual age model.

$$v_i = \begin{cases} v_{i-1} + (1 - a_{CM_i})x_i \text{ KIJIMA} - \text{I If CM is performed} \\ (1 - a_{CM_i})(v_{i-1} + x_i)\text{KIJIMA} - \text{II If PM is performed} \end{cases} \tag{2.13}$$

$$a_{\mathrm{CM}_i} = RF_{\mathrm{CM}_i}, a_{\mathrm{PM}_i} = RF_{\mathrm{PM}_i}$$

Values of virtual age 'v_i' for corresponding PM_j and CM_k are obtained from virtual age model, as tabulated in Table 2.4.

The time to failure (TTF) is obtained from Table 2.4, as shown Equation (2.14). To get the failure distribution, failure time prior to failure (CM) is calculated using Equation (2.14), as tabulated as Table 2.5.

These data points are now fitted in different distributions, and the most suitable distribution is selected. The process is shown in Figure 2.2. Finally, the parameters of the distribution are obtained.

$$\mathrm{TTF}_i = t_{v_i} = v_{i-1} + x_i, \text{ when } i \in \mathrm{CM} \qquad (2.14)$$

Here, t_{v_i} is the virtual age before the restoration at i^{th} maintenance action (CM/PM) and rest have their usual meanings.

Table 2.4 Virtual age from failure and maintenance time using calculated RF

Time to maintain	x_i in months	RF	Maintenance type	Virtual age
t_1	x_1	$RF(PM_1)$	TOH	v_1
t_2	x_2	$RF(CM_1)$	CM	v_2
t_3	x_3	$RF(PM_1)$	TOH	v_3
t_4	x_4	$RF(PM_2)$	IOH	v_4
t_5	x_5	$RF(PM_1)$	TOH	v_5
t_6	x_6	$RF(CM_2)$	CM	v_6
t_7	x_7	$RF(PM_1)$	TOH	v_7
t_8	x_8	$RF(CM_3)$	CM	v_8
t_9	x_9	$RF(CM_4)$	CM	v_9
t_{10}	x_{10}	$RF(CM_5)$	CM	v_{10}
t_{11}	x_{11}	$RF(CM_6)$	CM	v_{11}
t_{12}	x_{12}	$RF(PM_3)$	POH	v_{12}
t_{13}	x_{13}	$RF(PM_1)$	TOH	v_{13}
t_{14}	x_{14}	$RF(CM_7)$	CM	v_{14}
t_{15}	x_{15}	$RF(PM_1)$	TOH	v_{15}
t_{16}	x_{16}	$RF(CM_8)$	CM	v_{15}

Table 2.5 TTF points

t_{v2}	t_{v6}	t_{v8}	t_{v9}	t_{v10}	t_{v11}	t_{v14}	t_{v16}

t_{V2} t_{V6} t_{V8} t_{V9} t_{V10} t_{V11} t_{V14} t_{V16}

These virtual age data fitted into distribution to determine the parameters of the suitable distribution.

Distribution parameters e.g. Weibull (Shape parameter, scale parameter, etc.)

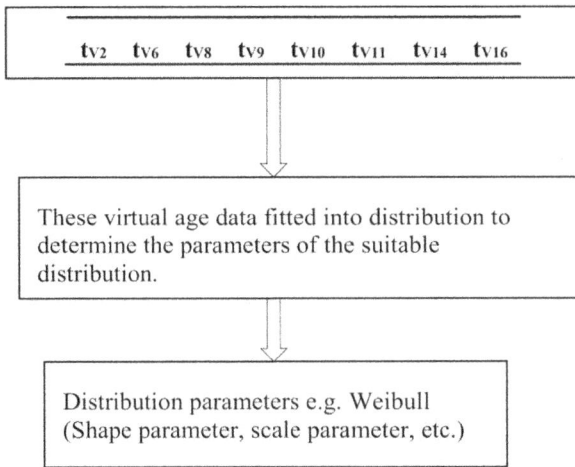

Figure 2.2 Process for obtaining failure distribution parameters.

2.3 CASE STUDY

Traction motor used in electric locomotive is considered as the RS for this study. Traction motors are used to power the wheelset of electrically operated rail vehicles. The failure data of the traction motor has been collected, as given in Table 2.6.

Table 2.6 Failure and maintenance data of TM

Type of maintenance	Time to maintenance (months)
TOH	18
CM	33
TOH	36
IOH	54
TOH	72
CM	78
TOH	90
CM	93
CM	99
CM	106
POH	108
TOH	126
CM	141
TOH	144
CM	147

In this study, three types of PM have been considered: term overhaul (TOH), intermediate overhaul (IOH) and periodic overhaul (POH). So, there are four types of maintenance (three PM and one CM) in total. In this case, the first time to failure (FTTF) distribution is considered as Weibull distribution. So, there are six parameters: two parameters for Weibull distribution (β, θ); one from CM (RF of CM a_{CM}) and three from PM (RF of three PM (a_{TOH}, a_{IOH}, a_{POH}). The system is briefly described in Table 2.6 along with its maintenance policy. In this case study, major overhauls of the traction motor are considered. Minor overhauls may be considered for more accurate virtual age calculation, which will give more accurate parameters of the Weibull distribution. Minor overhauls and RF due to cleaning and lubrication are not considered in this study as they have very low impact on the RF compared to replacement of the component. RF due to lubrication and cleaning needs more study to obtain its impact on the restoration of the component.

2.3.1 System description

Traction motor is one key component of the electric locomotive that delivers driving power to the wheel. Its proper maintenance is essential for the good reliability and availability of electric locomotive in service. It is a six-pole DC series wound motor with commutating poles. It has a forced ventilation system, and cooling air is supplied by the separate motor blower. It is attached to its respective driving axle of bogie by nose suspension, and tractive effort is transmitted from the traction motor to the axle through gear device with a single reduction. In this traction motor, axle taper roller suspension drive (bearing) is used. As per maintenance policy, there are three major overhauls: TOH, IOH and POH. The interval for each overhaul is different for different types of locomotives in which the traction motor is fitted. In our case, TOH, IOH and POH are performed at every 18, 54 and 108 months, respectively. In each overhaul, there is a set of activities described in maintenance manuals. Cleaning, lubrication, greasing and replacement of components are specified in the manuals. Based on the maintenance activities and component replacement, different levels of restoration are achieved during different overhaul activities.

2.3.2 Parameter estimation using MLE

First, we are presenting the traditional approach used for RS analysis taking RFs for each maintenance action and estimating them along with distribution parameters for real problem. Failure and maintenance data of the traction motor have been collected and shown in Table 2.6. From Table 2.6 and the use of maintenance policy, a likelihood function has been obtained. MATRix LABoratory (MATLAB®) software is used for coding an algorithm used for parameter estimation. The MLE equation has been optimized using optimization tools available in MATLAB® and its results are summarized in Table 2.7.

Table 2.7 Parameters obtained using MLE approach

Tool	θ	β	a_{CM}	a_{TOH}	a_{IOH}	a_{POH}
GA Parameters	19.1	1.11	0.97	0.72	0.91	1
LB	18	1.1	0.1	0.1	0.1	0.1
UB	100	5	1	1	1	1
Parameters	13.7	1	0.99	0.9	0.82	0.98
LB	10	0.8	0.1	0.1	0.1	0.1
UB	100	5	1	1	1	1
Parameters	7.39	2.4	0.98	0.73	0.21	0.97
	4.28	1.65	1	0.43	0.43	0.97
	2.02	1.02	1	0.88	0.67	0.99
	12.2	2.19	0.97	0.35	0.93	0.96
	10	1.88	0.96	0.92	0.68	0.99
LB	1	0.1	0.1	0.1	0.1	0.1
UB	100	5	1	1	1	1
Fsolve parameters	9.99	1.5	0.24	0.5	0.6	0.52
Starting point	10	1.2	0.4	0.5	0.6	0.8
Parameters	15	1.33	0.17	0.5	0.6	0.29
Starting point	15	0.8	0.4	0.5	0.6	0.8
Parameters	15	1.25	0.37	0.6	0.6	0.06
Starting point	15	0.8	0.5	0.6	0.6	0.4

All the six parameters have been estimated using global optimization techniques such as genetic algorithm (GA) and *function fsolve* in MATLAB® *Function fsolve* solves systems of non-linear equations of several variables. *Function fsolve* implements three different algorithms: trust region dogleg, trust region reflective and Levenberg–Marquardt. *Function fsolve* is sensitive to the initial estimation provided by us. Having an idea of what answers we expect for the variables, we choose initial guess. We should know the number of variables we are solving for and the number of equations used should match the number of variables. Values of the parameters vary for different upper and lower bounds chosen for the parameters.

From Table 2.7, it is evident that in GA for different lower bounds (LB) and upper bounds (UB), different values of the parameter are obtained. Similarly, for in function *fsolve* for different starting points, different values for the parameters are obtained. Therefore, it is very difficult to obtain consistent values of the parameter using MLE for parameters more than 3 (say 6).

Virtual age at each failure and maintenance time can be calculated, and the results are shown in Table 2.8. From the virtual age points, failure times can be obtained. TTF calculation is shown in Table 2.9. These data are used

Table 2.8 Calculation of virtual age using MLE method

Time to maintain	X_i in months	RF	Maintenance type	Virtual age
18	18	0.35	TOH	16.17
33	15	0.97	CM	16.62
36	3	0.35	TOH	18.57
54	18	0.93	IOH	19.83
72	18	0.35	TOH	31.53
78	6	0.97	CM	31.71
90	12	0.35	TOH	39.51
93	3	0.97	CM	39.60
99	6	0.97	CM	39.78
106	7	0.97	CM	39.99
108	2	0.96	POH	40.07
126	18	0.35	TOH	51.77
141	15	0.97	CM	52.22
144	3	0.35	TOH	54.17
147	3	0.97	CM	54.26
θ	12.18			
β	2.19			

Table 2.9 TTF points using MLE

TTF (months)	31.17	37.52	42.51	45.60	57.17	66.77

to find the Weibull distribution parameter (θ, β) assumed in MLE, using the Weibull probability plot, as shown in Figure 2.3 shown below.

From these results, it is concluded that there is inconsistency in parameter estimation using MLE for more than three parameters.

2.3.3 Parameter estimation using proposed approach

The parameter estimation using MLE approach for large number of parameters does not give very reliable results. So, a new methodology is proposed to determine the RF using AHP method. The method is illustrated in Table 2.10 for our considered system as traction motor.

Maintenance activities for each maintenance type are tabulated in Table 2.10. Here, Y stands for replacement and N represents non-replacement of the corresponding component.

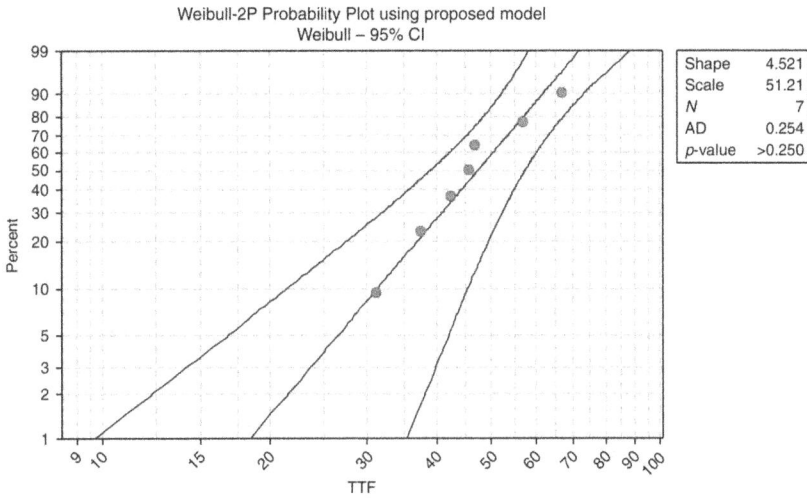

Weibull-2P Probability Plot using proposed model
Weibull – 95% CI

Shape	4.521
Scale	51.21
N	7
AD	0.254
p-value	>0.250

Figure 2.3 Weibull probability plot using parameters obtained from MLE.

Table 2.10 Replacement of components of different
types of maintenance of traction motor

Components of traction motor	TOH	IOH	POH
Stator	N	Y	Y
Rotor	N	Y	Y
Pinion	N	Y	Y
Terminal box	N	N	Y
Rotor bearing	Y	Y	Y
TM bellows	N	Y	Y
End plates	N	N	N
Temperature sensor	N	Y	Y
Speed sensor	N	N	N

2.3.3.1 Apply AHP to calculate weightage of each component

All three levels of AHP can be represented in a relationship diagram as Figure 2.4. Based on this, AHP is applied to calculate the weightage of the components on RF of the system using the proposed model.

First, the comparison matrix for different criteria is made. From the pairwise comparison matrix, normalized AHP matrix is formed to obtain the priority of each criterion. Priority vector for criteria is obtained as given in Table 2.11. From Table 2.11, it is clear that functionality has the highest

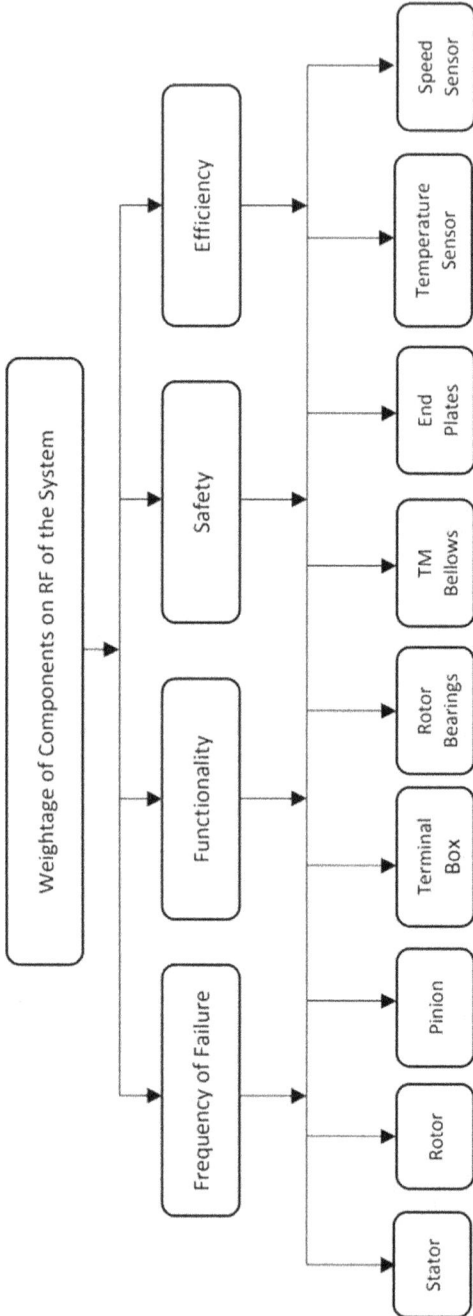

Figure 2.4 Levels of AHP applied for TM.

Table 2.11 Priority vector of criteria

Criteria	Weight (C_i)
Frequency of failure (FoF)	0.24
Functionality	0.60
Safety	0.10
Efficiency	0.05
Sum	1.00

Table 2.12 Priority vector of components with respect to frequency of failure

Component	Weights ($W_{i,j}$)				Final weight
	FoF (0.24)	Functionality (0.60)	Safety (0.10)	Efficiency (0.05)	$W_j = \sum_{i=1}^{4} W_{i,j} \times C_i$
Stator	0.15	0.21	0.06	0.20	0.179
Rotor	0.12	0.21	0.06	0.20	0.171
Pinion	0.18	0.16	0.09	0.09	0.153
Terminal box	0.09	0.12	0.09	0.09	0.111
Rotor bearing	0.32	0.12	0.14	0.08	0.170
TM bellows	0.02	0.07	0.14	0.17	0.068
End plates	0.02	0.07	0.15	0.02	0.062
Temperature sensor	0.04	0.02	0.14	0.08	0.044
Speed sensor	0.04	0.02	0.13	0.08	0.042
Sum	**1.00**	**1.00**	**1.00**	**1.00**	**1.00**

weightage among all the criteria. After determining priority vector for criteria, priority vector for alternatives is obtained with respect to each criterion. Priority vector and matrix for components with respect to each criterion are given in Table 2.12. For each criterion, there is a different component, which has the maximum weightage. In the case of frequency of failure, a rotor bearing has the maximum weightage, as it fails rapidly compared to other components. With respect to functionality, both rotor and stator have the maximum weightage. In the case of safety criteria, end plates, TM bellows, temperature sensor and rotor bearings have similar weightages, greater than other components. With respect to efficiency, stator and rotor have the maximum weightage followed by rotor bearings.

2.3.3.2 RF calculation for different types of PM and CM

Here, 'Y' is replaced with '1' and 'N' is replaced with '0.' 1 stands for renewal, it says that when an item is replaced, it is renewed, and the amount of restoration it gives the system is determined by the product of 1 and its weightage

Table 2.13 RF for TOH, IOH and POH of traction motor

RF (traction motor)	Replacement			Final weight (W_j)
	TOH ($R_{1,j}$)	IOH ($R_{2,j}$)	POH ($R_{3,j}$)	
Stator	0	1	1	0.179
Rotor	0	1	1	0.171
Pinion	0	1	1	0.153
Terminal box	0	0	1	0.111
Rotor bearing	1	1	1	0.170
TM bellows	0	1	1	0.068
End plates	0	0	0	0.062
Temperature sensor	0	1	1	0.044
Speed sensor	0	0	0	0.042
Restoration factor	**0.170**	**0.785**	**0.896**	
$$\left(RF_i = \sum_{j=1}^{9} R_{i,j} \times W_j \right)$$				

on RF of the system by using AHP. Similarly, 0 stands for ABAO condition, and its impact on restoration will be simply zero. RF for each maintenance type is calculated corresponding to maintenance type, as given in Table 2.13.

CM is performed after occurrence of failure during operation. According to the proposed model, its RF must depend on the maintenance activity performed during CM. If failure occurs due to any component, it can be repaired by the replacement of that component. To calculate RF in this case, it is the sum of weightage of the replaced items multiplied by 1. RFs for three CM are calculated and shown in Table 2.14.

Table 2.14 Restoration factor for different CM of traction motor

RF (traction motor)	CM1 ($R_{1,j}$)	CM2 ($R_{2,j}$)	CM3 ($R_{3,j}$)	Final weight (W_j)
Stator	0	0	0	0.179
Rotor	1	0	0	0.171
Pinion	0	1	0	0.153
Terminal box	0	0	0	0.111
Rotor bearing	1	1	1	0.170
TM bellows	1	0	1	0.068
End plates	0	0	0	0.062
Temperature sensor	0	0	0	0.044
Speed sensor	0	1	0	0.042
Restoration factor	**0.240**	**0.365**	**0.238**	
$$\left(RF_i = \sum_{j=1}^{9} R_{i,j} \times W_j \right)$$				

2.3.3.3 Calculating failure distribution parameters

After RF calculation, failure distribution parameters of the system are determined. According to the proposed model, virtual age is first determined at all the points of failure (CM) and PM by applying the values of RF for different types of PM and CM, as shown in Table 2.15.

TTF points corresponding to failure (CM) are calculated. Failure data points are tabulated in Table 2.16. Virtual age data are treated as failure data and fitted into the Weibull probability plot, as shown in Figure 2.5. Anderson-Darling (A-D) test for goodness of fit (GOF) value is 0.365, and p-value greater than 0.250 shows good fitness. The parameters of the Weibull distribution, given in Table 2.17, are shape parameter 6.00 and scale factor 38.00 months. Mean time to failure (MTTF) of TM is 35.226 months. Reliability vs. Time plot and Real age vs. Virtual age plot are also shown in Figure 2.6 and Figure 2.7, respectively, to illustrate the effect of restoration due to PM and CM action on TM. Finally, all the RF and failure distribution parameters are obtained using the proposed model.

Table 2.15 Virtual age of TM at different points obtained using proposed model

Time to maintain (months)	Xi or TBF (months)	RF	Maintenance type	Virtual age
18	18	0.17	TOH	14.94
33	15	0.26	CM1	26.11
36	3	0.17	TOH	24.16
54	18	0.79	IOH	9.06
72	18	0.17	TOH	22.46
78	6	0.17	CM2	27.43
90	12	0.17	TOH	32.73
93	3	0.18	CM3	35.20
99	6	0.26	CM5	39.66
106	7	0.29	CM6	44.64
108	2	0.90	POH	4.87
126	18	0.17	TOH	18.98
141	15	0.17	CM7	31.41
144	3	0.17	TOH	28.56
147	3	0.32	CM8	30.59

Table 2.16 TTF points estimated using proposed model

TTF (months)	28.46	29.94	31.56	33.98	35.73	41.20	46.66

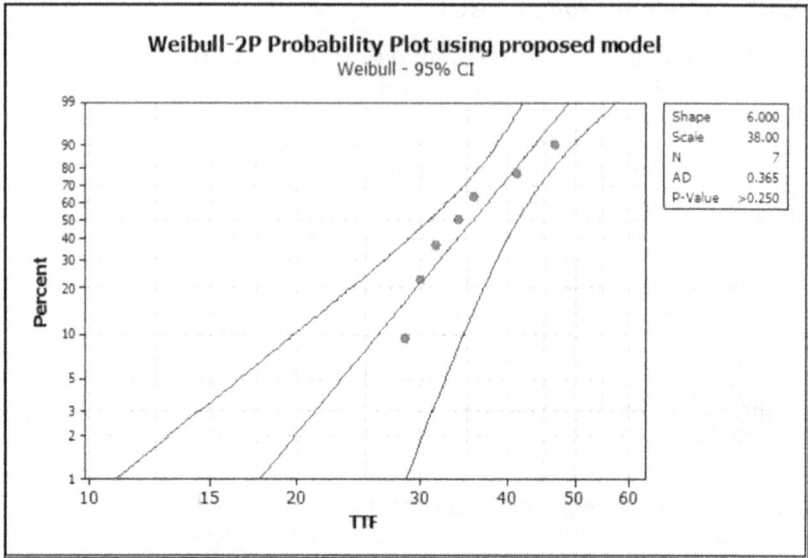

Figure 2.5 Weibull probability plot using proposed model.

Table 2.17 Failure distribution parameters of traction motor

Distribution	Beta	Theta (months)
Weibull 2-P	6.00	38.00

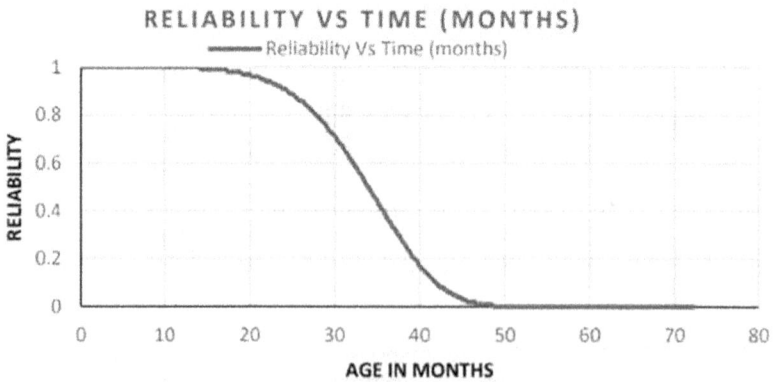

Figure 2.6 Reliability vs. Time plot of traction motor.

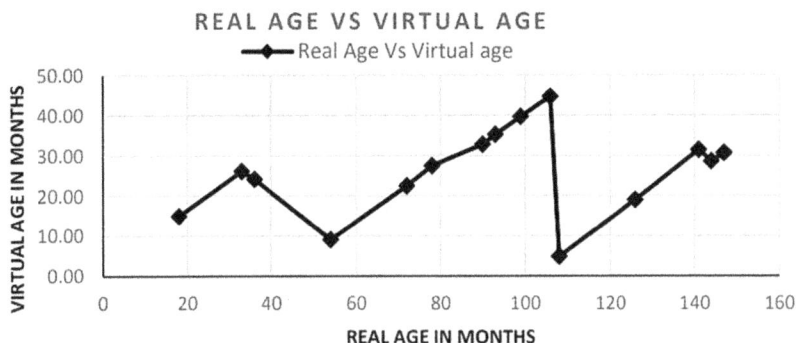

Figure 2.7 Real age vs. Virtual age of traction motor (Impact of maintenance).

2.3.3.4 Validation

Finally, all the RF and failure distribution parameters are obtained using the proposed model. The proposed model to predict RF for different maintenance types can be validated by comparing number of failure (NOF) estimated from the proposed model and actual NOF. NOF calculated is given in Table 2.18. NOF estimated from the proposed model is 8.63 and actual NOF is 7. The result will be more accurate if we consider the minor

Table 2.18 Number of failure calculation using the parameters from the proposed model

Time to maintain (Months)	Xi or TBF (Months)	RF	Maintenance type	Virtual age	Conditional NOF
18	18	0.17	TOH	14.94	0.011
33	15	0.26	CM1	26.11	0.236
36	3	0.17	TOH	24.16	0.097
54	18	0.79	IOH	9.06	1.799
72	18	0.17	TOH	22.46	0.130
78	6	0.17	CM2	27.43	0.134
90	12	0.17	TOH	32.73	1.107
93	3	0.18	CM3	35.20	0.283
99	6	0.26	CM5	39.66	0.992
106	7	0.29	CM6	44.64	2.135
108	2	0.90	POH	4.87	0.790
126	18	0.17	TOH	18.98	0.047
141	15	0.17	CM7	31.41	0.496
144	3	0.17	TOH	28.56	0.232
147	3	0.32	CM8	30.59	0.148
	Actual NOF	**7.000**		**Total NOF**	**8.637**

OH performed on TM. There is a good chance that multiple failures are recorded as single failure during single trip, which causes difference in estimated NOF and actual NOF. Overall, the estimated result obtained from the model is conservative, which is acceptable, considering the small number of data points.

2.4 CONCLUSION

Traction motor used in Indian railways is considered for the study which uses application of AHP to calculate the weightage of each component on system restoration. RF is calculated for CM and PM types by the proposed model and taking weightage of the component and maintenance action performed on the components. Virtual age of the system is determined by using RF for different maintenance types. Virtual age at the time of failure is taken as time to failure and fitted into Weibull distribution to obtain shape and scale parameters.

The traction motor has three major overhauls (TOH, IOH and POH) along with three types of minor PM. Only major overhauls are considered because the RF is considerable in these cases. RF obtained for TOH is 0.170, for IOH 0.785 and for POH 0.896. In the case of CM, RF is also calculated, which depends upon the components replaced during CM. Then, RF for all the CM types is calculated. The calculated RF is used to get the virtual age at each CM or PM points. Finally, according to the model, failure distribution parameters of Weibull distribution are calculated as $\beta = 6.00$ and $\theta = 38.00$ months.

In this work, only RF due to replacement is considered. RF due to cleaning and lubrication is not considered, as they have very little impact as compared to RF due to replacement. Decrease in the ability to restore the system due to aging of the system can be addressed in future work. Overall, this approach is easier and gives different RFs for different types of PM and CM. Using these RF values, failure distribution parameters are now easily obtained, instead of only one RF for different types of PM, as taken by usual approaches in literature.

REFERENCES

Doyen, Laurent, and Olivier Gaudoin. 2004. "Classes of Imperfect Repair Models Based on Reduction of Failure Intensity or Virtual Age." *Reliability Engineering and System Safety* 84 (1): 45–56. https://doi.org/10.1016/S0951-8320(03)00173-X.

Malik, Mazhar Ali Khan. 1979. "Reliable Preventive Maintenance Scheduling." *AIIE Transactions* 11 (3): 221–28. https://doi.org/10.1080/05695557908974463.

Meykar, Orest A. 1967. "Definitions of Effectiveness Terms: A Report on the Purpose and Contents of MIL-STD-721B." *IEEE Transactions on Aerospace and Electronic Systems* AES-3 (2): 165–70. https://doi.org/10.1109/TAES.1967.5408738.

Nakagawa, Toshio. 1979. "'Imperfect Preventive-Maintenance,' no. 5: 9529.1979b. 'Optimum Policies When Preventive Maintenance Is Imperfect'". *IEEE Transactions on Reliability* R-28 (4): 331–32. https://doi.org/10.1109/TR.1979.5220624.

Nasr, Arwa, Soufiane Gasmi, and Mounir Sayadi. 2013. "Estimation of the Parameters for a Complex Repairable System with Preventive and Corrective Maintena nce." *2013 International Conference on Electrical Engineering and Software Applications, ICEESA 2013.* https://doi.org/10.1109/ICEESA.2013.6578455.

Pham, Hoang, and Hongzhou Wang. 1996. "Imperfect Maintenance." *European Journal of Operational Research* 94 (3): 425–38. https://doi.org/10.1016/S0377-2217(96)00099-9.

Said, Uthman, and Sharareh Taghipour. 2016. "Modeling Failure and Maintenance Effects of a System Subject to Multiple Preventive Maintenance Types." *Proceedings - Annual Reliability and Maintainability Symposium* 2016-April. https://doi.org/10.1109/RAMS.2016.7448007.

Triantaphyllou, Evangelos, Boris Kovalerchuk, Lawrence Mann, and Gerald M. Knapp. 1997. "Determining the Most Important Criteria in Maintenance Decision Making." *Journal of Quality in Maintenance Engineering* 3 (1): 16–28. https://doi.org/10.1108/13552519710161517.

Trust, Applied Probability. 1989. "Some Results for Repairable Systems with General Repair Author (s): Masaaki Kijima" *Journal of Applied Probability* 26 (1): 89–102.

Yaez, Medardo, Francisco Joglar, and Mohammad Modarres. 2002. "Generalized Renewal Process for Analysis of Repairable Systems with Limited Failure Experience." *Reliability Engineering and System Safety* 77 (2): 167–80. https://doi.org/10.1016/S0951-8320(02)00044-3.

Yu, Pingjian, Joon Jin Song, and C. Richard Cassady. 2008. "Parameter Estimation for a Repairable System under Imperfect Maintenance." *Proceedings - Annual Reliability and Maintainability Symposium* (4): 428–33. https://doi.org/10.1109/RAMS.2008.4925834.

Chapter 3

Measurement and modeling of cutting tool temperature during dry turning operation of DSS

P. Kumar
National Institute of Technology Kurukshetra, Kurukshetra, India

O.P. Yadav
North Dakota State University, Fargo, ND, USA

CONTENTS

3.1 INTRODUCTION

Metal cutting is a standard process in manufacturing industries. During metal cutting, the temperature in the cutting zone increases. This temperature rise is the result of friction that occurs between interfaces of tool-chip and tool-workpiece. However, the primary cause is the plastic deformation of the material. Consequently, the energy of any conventional metal cutting process is converted into heat. Hence, it results in a rise in temperature within a cutting zone (Veiga et al. 2021, Bhattacharya et al. 2021). This rise in cutting zone temperature affects machining characteristics. Therefore, it becomes essential to have certain information about heat generation and temperature increase during the machining process (Knight and Boothroyd 2005, Stephenson and Agapiou 2018). Augmentation in maximum temperature at cutting tool's clearance and rake face accelerates tool wear. Also, metallurgical changes may take place in tool or workpiece material due to maximum temperature and gradient. Also, some physical and chemical reactions may occur at this higher temperature induced during the metal cutting process, which ultimately results in rapid tool wear, increase in power consumption and deterioration of surface finish of the product, and dimensional accuracy (Nedić and Erić 2014, Pal, Choudhury, and Chinchanikar 2014, Valera and Bhavsar 2014, Kumar and Misra 2019). Furthermore, the magnitude of cutting tool temperature and heat distribution between cutting tool, chip, and workpiece material is due to their respective material's mechanical and chemical properties, machining regime, cutting tool geometry, cutting environment, and other parameters (Aneiro, Coelho, and Brandão 2008, Thakare and Nordgren 2015, Hosseini et al. 2014, Chinchanikar and Choudhury 2014). So, it becomes necessary to find out the temperature in the metal cutting zone and also predict it if possible.

Since many phenomena (mechanical, thermal, chemical, etc.) occur between the cutting tool and workpiece point of contact, the measurement of temperature in metal cutting operations becomes quite a tricky job. Hence, model development for predicting temperature for the turning or any machining process is a tremendously tough process. As a result, temperature prediction of the machining process with accuracy and repeatability becomes a challenge (Abukhshim, Mativenga, and Sheikh 2005). Additionally, with a shortage of adequate experimental data it's challenging to develop and verify a mathematical model. However, plentiful attempts have been made for the measurement of machining zone temperature (Abukhshim, Mativenga, and Sheikh 2006). Equation (3.1) represents an analytical model developed by Lowen and Shaw (Loewen and Shaw 1954) to measure cutting temperature during machining. They formulated a relationship between cutting temperature, cutting speed, and feed rate.

$$\theta_t = V^{0.5} \times f^{0.3} \qquad\qquad (3.1)$$

Here, θ_t = average cutting temperature, V = cutting speed, and f = feed rate.

In 1998, for the specific range of feed rate and cutting speed, Chu and Wallbank (1998) developed a relation between cutting parameters and cutting temperature. Kus et al. (2015) performed a comparative study using thermocouple and infrared (IR) pyrometer for tool-chip interface temperature measurement during dry machining of AISI 4140 alloy. They concluded that with an increase in machining regimes, there is a rise in temperature. They concluded that cutting speed is the vital parameter affecting the temperature, whereas other factors like feed rate and depth of cut were not significant. Shihab et al. (2014) did the turning of AISI hard alloy steel and suggested an RSM model using statistical analysis. FEM model was also developed and compared by researchers based on experimental data (Kus et al. 2015, Kryzhanivskyy et al. 2015).

Higher cutting zone temperature is observed in the machining process, which ultimately accelerates the cutting tool wear through various wear mechanisms, namely abrasion, diffusion, and adhesion (Chinchanikar, Choudhury, and Kulkarni 2013). There is a need for mathematical model development and evaluation to predict cutting tool temperature (CTT). Hence, a predictive model considering machining regimes for a particular tool and work material combination becomes enormously precious. Özel (2009) developed a 3D finite element model to investigate the effect of variable edge micro-geometry PCBN cutting tool on the cutting temperature for AISI 4340 turning operation. In another study, Fnides, Yallese, and Aouici (2008) performed hard turning of AISI H11 without coolant, which was performed to investigate the effect of cutting zone temperature. The various parameters considered were cutting speed, feed rate, and depth of cut, and machining was performed using a mixed ceramic insert (CC650). Their investigation found that cutting speed has the main influence on the cutting zone temperature rather than the feed rate and depth of cut. Lin et al. (2008) used CBN tools to turn AISI 4340 alloy and investigated the effect of cutting speed on cutting temperature. They observed the direct relationship between the temperature and the cutting speed.

Bouchelaghem et al. (2010), in their experimental investigation, found the directly proportional relation between cutting temperature and cutting parameters. They also concluded that prolonged machining time leads to tool wear, which ultimately results in cutting zone temperature. For their experimental study, they selected AISI D3 (60 HRC) as workpiece material, and the turning was performed using CBN as a cutting tool material. Aouici et al. (2010) did perform turning under the dry machining environment of AISI H11, and cutting parameters on cutting temperature were explored. Using the RSM technique of DOE, they developed a model between cutting temperature and machining parameters. In their study, they observed a directly proportional relationship between the temperature rise and the machining variables. Amritkar et al. (2012) designed and developed an economical setup to measure the CTT. They related the induced electromotive forces (emf) with the cutting temperature produced during machining. SAE 8620 was machined using an uncoated WC tool at various cutting speed and feed rates for their study. They further modeled the experimental outcome using regression analysis. They also tested the setup performance with other work materials such as EN19, EN31, mild steel, and SS 304 while utilizing the same cutting tool grade. In their investigation, they observed the accuracy and repeatability of results. Lin, Lee, and Weng (1992), in their experimental study, measured the tool-chip temperature using an IR pyrometer for the high-speed machining (600 m/min) of AISI 1045 using carbide and ceramic tools. Similarly, during high-speed machining of H13 hardened steel Dewes et al. (1999) used an IR camera and thermocouple approach to evaluate tool-chip temperature. Young (1996) utilized the IR camera during orthogonal cutting of AISI 1045 steel and measured the interface and the chip temperature of the back section.

In heavy sectors such as oil and gas, nuclear power, and chemical processing, Duplex Stainless Steel (DSS) is in great demand (Gowthaman, Jeyakumar, and Saravanan 2020). The most extensively used duplex stainless steel is 2205, which has enormous corrosion resistance and high strength. The various components like pressure vessels, tanks, piping, tubing, heat exchangers, digesters, and bleaching equipment are made with DSS. Various researchers so far have tried many methods and techniques to determine the temperature of the cutting tool or machining zone. Numerous efforts have also been made to estimate CTT and, for the parametric optimization of machining parameters, to minimize the machining zone temperature. However, no experimental study or cutting temperature model for dry turning of DSS using coated carbide inserts has been reported so far. In this research work, an attempt is to cover this gap. A relationship between cutting temperature and influential machining parameters have been proposed using experimental results obtained during conventional turning of DSS (2205), which have a wide application range in various engineering fields, namely pulp and paper, oil and gas, chemical industry, marine industry etc., due to its superior properties and cost-effectiveness.

3.2 MATERIALS AND METHODS

Trial runs in a dry cutting machining environment are carried out to examine the influence of the machining regime on CTT. The experiment is carried out using L_9 OA of Taguchi's technique to assess the impact of individual machining regimes. The various levels of machining parameters and their levels considered in this experimentation are shown in Table 3.1.

As depicted in Figure 3.1, on a traditional machine tool OKUMA lathe (*Make: Japan*), turning trials runs were carried out.

The supplier provided the workpiece, and its elemental composition was confirmed via an EDS test. Figure 3.2 depicts the results of the EDS test.

The rod was prepared for final experimentation by removing surface material to a certain depth. During experiments, between the three-jaw chuck and the lathe's tailstock, a workpiece of about 30 cm length was held. The machining was done with coated "Widia" indexable inserts with a 0.4 mm nose radius, which have TNMA form identification as per ISO standards.

Table 3.1 Cutting parameters and levels

Cutting parameters	Symbols	Unit	Parameter levels		
Cutting speed	s	m/min	39.77	55.29	73.47
Feed	f	mm/rev	0.07	0.14	0.21
Depth of cut	d	mm	0.2	0.4	0.6

Figure 3.1 Lathe.

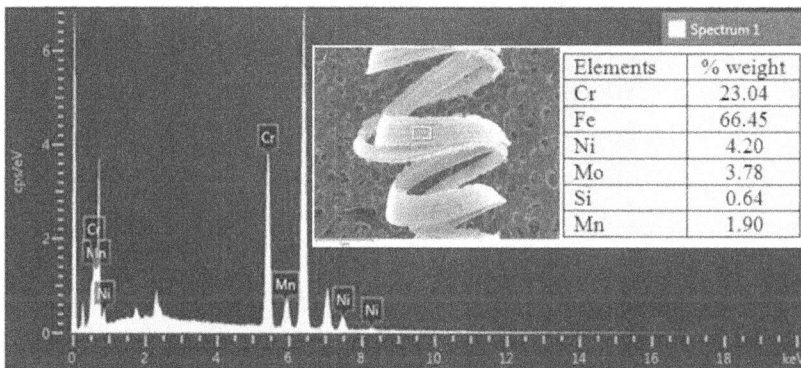

Figure 3.2 EDS of workpiece material.

The inserts were mechanically fastened on a robust tool holder of the WTJNL 2020 K16 type made by "TaeguTec." An infrared thermometer (IT) was used to measure the CTT, having a temperature measuring range from -25^0C to 1200^0C. The instrument so used is shown in Figure 3.3.

This method was used due to its many advantageous features like a fast response, no adverse effect on work material, and easy to operate as it is portable (non-contact type), making it suitable for the present study. In this temperature measurement method, the emitted thermal energy from the surface of the body is utilized, giving information about the temperature of the body under investigation. However, this temperature measurement technique faces problems with exact surface emissivity from the body (under investigation) for the temperature measurement. The principle behind the IR

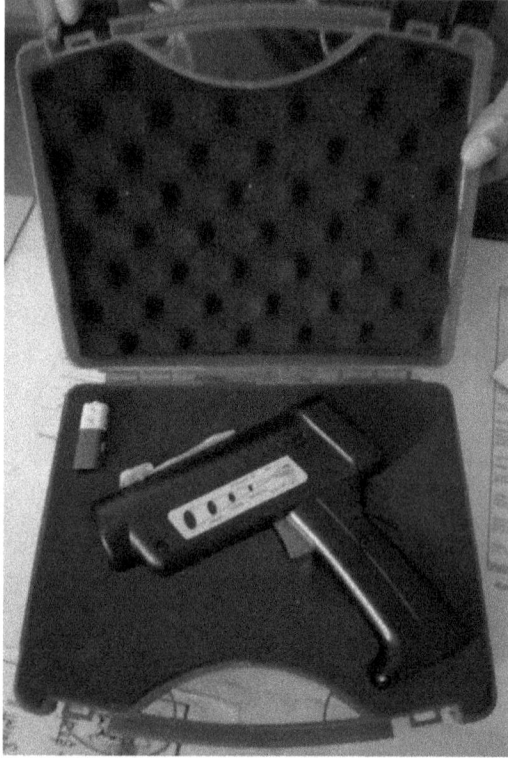

Figure 3.3 Infrared thermometer.

thermography is as follows, as shown in Equation (3.2). It relates the radiated energy (E) and the temperature of body (T) under investigation, and this relation is defined by Stefan Boltzmann law (Svetlitza et al. 2014).

$$E = \sigma \varepsilon T^4 \tag{3.2}$$

Here, $\sigma = 5.67 \times 10^{-8}$, T is the temperature in K, and ε is the emissivity coefficient of the body surface.

To measure the temperature of cutting tool, an IT was focused on the rake face of the insert. All the measured values of temperatures are summarized in Table 3.2. The Design of Experiment (DOE) method was implemented for experimentation to save resources (time and cost). Moreover, DOE produces almost equivalent results so produced by the One-Factor-At-A-Time (OFAT) approach without loss of information. Taguchi techniques have been successfully implemented for complex machining problems (Pawan and Misra 2018). The selected OA has eight degrees of freedom, which suits for our parametric space. Hence, the effect of all machining parameters can be easily and independently examined by doing just nine experiments. But it lacks

Table 3.2 Experimental results

Sr. No.	s (m/min)	f (mm/rev)	d (mm)	CTT (°C)
1	39.77	0.07	0.2	41
2	39.77	0.14	0.4	57
3	39.77	0.21	0.6	64
4	55.29	0.07	0.4	59
5	55.29	0.14	0.6	77
6	55.29	0.21	0.2	58
7	73.47	0.07	0.6	180
8	73.47	0.14	0.2	79
9	73.47	0.21	0.4	89

the information related to the interaction effect of machining parameters. Taguchi method, which also ensures the optimum condition, helps to observe the influence of machining regimes on the response (Kumar and Misra 2020). Table 3.2 is depicting Taguchi's L_9 OA with its output as a CTT. According to this design matrix, the cutting tests are performed in a random order to avoid parameter bias during experimentation. A correlation between the machining regimes and CTT was developed by varying the machining regimes as per the experimental plan given by OA design matrix.

3.3 RESULTS AND DISCUSSION

Turning trial runs were performed according to the L_9 design matrix. For each trial run, a fresh insert is used, and the temperature of the rake face of the cutting tool was recorded with the help of IT. The experimental observations are summarized in Table 3.2. All runs were performed in randomized order. This data was then examined using the MINITAB software. Moreover, the results of experimentation were further modeled mathematically for prediction purposes in the next section.

Figure 3.4 depicts the influence of machining regimes on the cutting tool temperature.

It is easily observable from Figure 3.4 that cutting speed (s) is the dominant parameter affecting the CTT. With the rise in cutting speed (s), there is an immediate increase in the temperature of the cutting tool. This temperature rise is attributed to an increase in friction and strain rate in the primary and the secondary deformation zone with increasing cutting speed (s). But feed (f) does not influence the rise in temperature of the cutting tool. Depth of cut (d) was also found quite influential in this case. At a lower depth of cut (d) value, there is less material to deform. However, at a higher depth of cut (d) setting, a large quantity of matter is deformed; hence, more energy is

Main Effects Plot for Cutting Tool Temperature

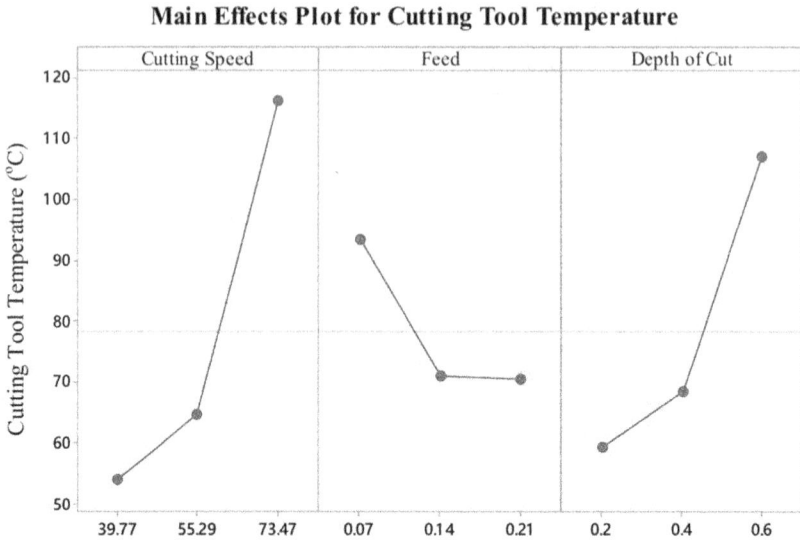

Figure 3.4 Main effect plot for cutting tool temperature.

Table 3.3 ANOVA results for CTT

Source	DF	Adj SS	Adj MS	f-value	p-value	% contribution
Model	3	10155.4	3385.1	5.39	0.050	
s	1	5953.7	5953.7	9.49	0.027	45
f	1	793.5	793.5	1.26	0.312	06
d	1	3408.2	3408.2	5.43	0.067	26
Error	5	3138.2	627.6			23
Total	8	13293.6				

required for plastic deformation of workpiece material (to convert it into chips), which subsequently increases in temperature of the cutting zone. The maximum CTT observed in this study was 180°C.

The relative relevance of the cutting parameter on the CTT is determined using ANOVA. The study is conducted at a statistically significant level of 95% ($\alpha = 0.05$). The results of ANOVA analysis are shown in Table 3.3. The relative impact of different factors on response was also calculated. From this analysis, it was concluded that cutting speed (s) is having the highest effect and contribution on the CTT rise. The second most affecting parameter is the depth of cut (d), and feed (f) has a negligible effect. But at the 95% statistical level, cutting speed (s) is the only one which was found to be significant.

3.4 EMPIRICAL MODELING

To describe the relationship between CTT and cutting parameters, regression analysis is used. From the experimentation so performed using coated tools to explore the effects of machining regimes on the temperature of cutting tool during dry turning of DSS, an empirical relationship is established, as shown by Equation (3.3.3).

$$CTT = -51.3 + 1.867 \times s - 164 \times f + 119.2 \times d \qquad (3.3)$$

The regression coefficients obtained from regression analysis are shown in Table 3.4. The mean change in the CTT for one unit of change in the machining regimes while holding other predictors in the model constant is represented by these regression coefficients.

The negative sign of a regression coefficient represents a negative correlation between each independent variable and dependent variable, and vice versa is also true. Figure 3.5 (a–b) showing the calculation results of the cutting tool's rake face temperature compared with the experimentally measured values. A good agreement was obtained (R^2 = 0.7639) between the analytical results and the experimental ones.

Figure 3.6 (a) is showing the normal plot of residuals. As can be observed, distinct data points are spread near the 45^0 line. It means that data points (residuals) are normally distributed and that they are independent. On the other hand, Figure 3.6 (b) represents the distribution of residuals over the fitted values. We can observe that there is no pattern or trend among different data points which makes analysis successful. This means the predictive model so developed can be used with certainty for prediction purposes (Gowthaman, Jeyakumar, and Saravanan 2020).

Table 3.4 Coefficients form regression analysis

Term	Coef.	SE Coef.	t-value	p-value	Variance inflation factor
Constant	−51.3	45.5	−1.13	0.310	
C1	1.867	0.606	3.08	0.027	1
C2	−164	146	−1.12	0.312	1
C3	119.2	51.1	2.33	0.067	1

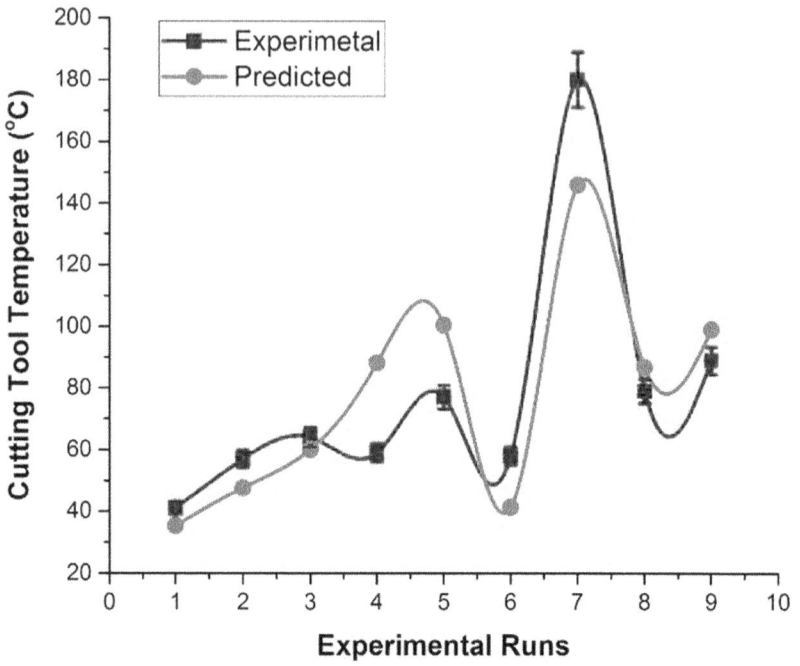

(a)

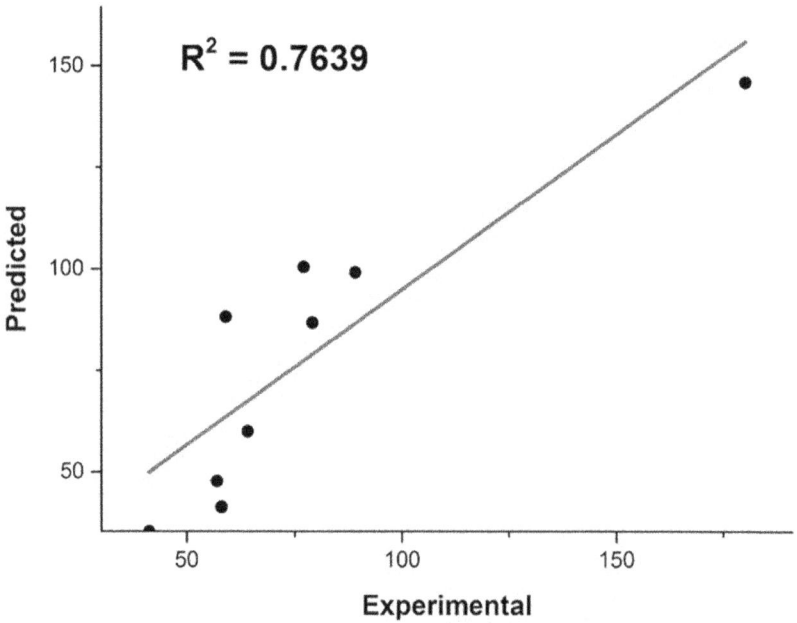

(b)

Figure 3.5 2D plots for (a) experimental versus predicted values (b) correlation.

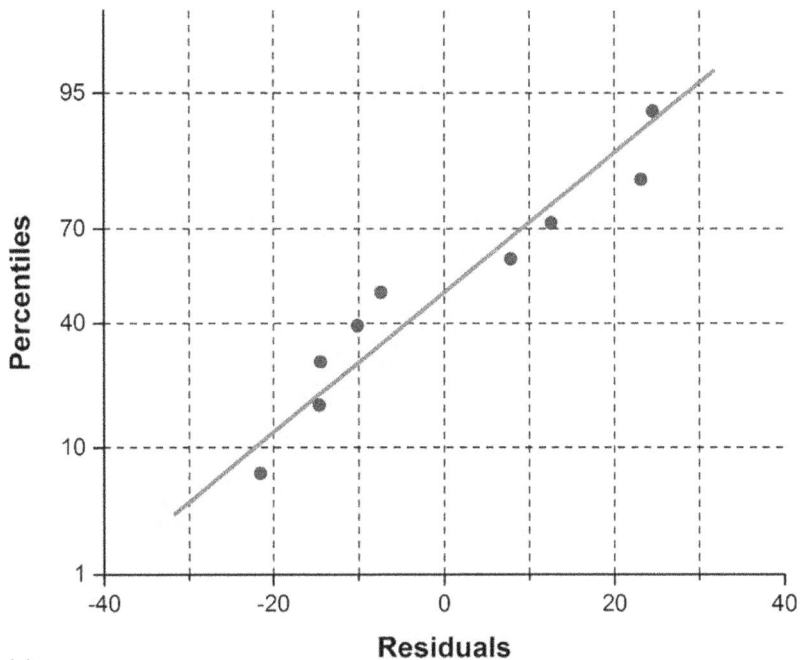

(a)

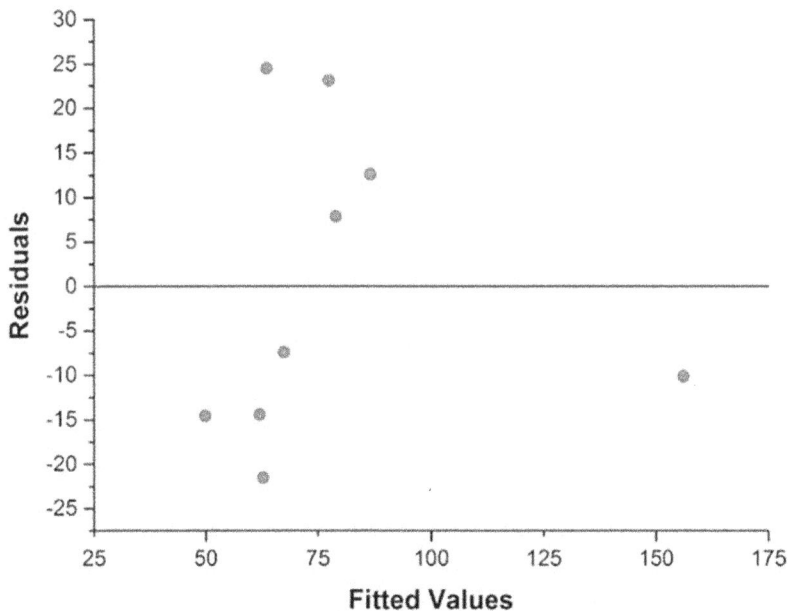

(b)

Figure 3.6 Residual results plots for (a) normality and (b) residual versus fits for CTT.

3.5 CONCLUSIONS

The main aim of this research work was on CTT measurement and modeling during dry turning of DSS (2205) on a conventional lathe using coated carbide inserts as a cutting tool. The effect of machining regimes on CTT was investigated, and a predictive model was developed for the studied parametric range. The cutting speed (s) found a critical factor for the rise in CTT. The influence of feed (f) was negligible, but the depth of cut (d) was also quite close to significance at a 95% statistical significance level. An empirical model with 76.39 % prediction efficiency is developed for the prediction of CTT as experimental results match well with the predicted CTT. Hence, good conformity was obtained between the analytical and experimental results.

The ANOVA study revealed that cutting speed (s) has the highest contribution (45%) on CTT followed by the depth of cut (d) with 26% contribution. Feed (f) has minimal influence on CTT rise (06% only), and its effect can be neglected. This experimental study's experimental and statistical approach presents a consistent method to the model present case. Additionally, this method can also be implemented to investigate and model other machining operations. Moreover, predictive models that can quantitatively predict the effect of machining regimes on the output characteristics are of great interest.

This study limits the interaction effect of machining regimes which might be crucial. Therefore, in future work, this work can be explored for the development of a better predictive model.

REFERENCES

Abukhshim, NA, PT Mativenga, and MA Sheikh. 2005. "Investigation of heat partition in high speed turning of high strength alloy steel." *International Journal of Machine Tools and Manufacture* 45 (15):1687–1695.

Abukhshim, NA, PT Mativenga, and Mohammed Aslam Sheikh. 2006. "Heat generation and temperature prediction in metal cutting: A review and implications for high speed machining." *International Journal of Machine Tools and Manufacture* 46 (7–8):782–800.

Amritkar, Abhijeet, Chandra Prakash, and Atul P Kulkarni. 2012. "Development of temperature measurement setup for machining." *World Journal of Science and Technology* 2 (4):15–19.

Aneiro, Federico M, Reginaldo T Coelho, and Lincoln C Brandão. 2008. "Turning hardened steel using coated carbide at high cutting speeds." *Journal of the Brazilian society of mechanical sciences and engineering* 30:104–109.

Aouici, H, MA Yallese, B Fnides, and Tarek Mabrouki. 2010. "Machinability investigation in hard turning of AISI H11 hot work steel with CBN tool." *Mechanics* 86 (6):71–77.

Bhattacharya, Shibaprasad, Partha Protim Das, Prasenjit Chatterjee, and Shankar Chakraborty. 2021. "Prediction of responses in a sustainable dry turning operation:

A comparative analysis." *Mathematical Problems in Engineering* 2021, Article ID 9967970.

Bouchelaghem, H, MA Yallese, Tarek Mabrouki, Abdelkrim Amirat, and Jean Francois Rigal. 2010. "Experimental investigation and performance analyses of CBN insert in hard turning of cold work tool steel (D3)." *Machining Science and Technology* 14 (4):471–501.

Chinchanikar, Satish, and SK Choudhury. 2014. "Evaluation of chip-tool interface temperature: effect of tool coating and cutting parameters during turning hardened AISI 4340 steel." *Procedia Materials Science* 6:996–1005.

Chinchanikar, Satish, SK Choudhury, and AP Kulkarni. 2013. "Investigation of chip-tool interface temperature during turning of hardened AISI 4340 alloy steel using multi-layer coated carbide inserts." *Advanced Materials Research* 701:354–358.

Chu, TH, and J Wallbank. 1998. "Determination of the temperature of a machined surface."

Dewes, RC, E Ng, KS Chua, PG Newton, and DK Aspinwall. 1999. "Temperature measurement when high speed machining hardened mould/die steel." *Journal of materials processing technology* 92:293–301.

Fnides, B, MA Yallese, and H Aouici. 2008. "Hard turning of hot work steel AISI H11: Evaluation of cutting pressures, resulting force and temperature." *Mechanics* 72 (4):59–63.

Gowthaman, PS, S Jeyakumar, and BA Saravanan. 2020. "Machinability and tool wear mechanism of Duplex stainless steel–a review." *Materials Today: Proceedings* 26:1423–1429.

Hosseini, SB, Tomas Beno, U Klement, J Kaminski, and K Ryttberg. 2014. "Cutting temperatures during hard turning—Measurements and effects on white layer formation in AISI 52100." *Journal of Materials Processing Technology* 214 (6):1293–1300.

Knight, Winston A, and Geoffrey Boothroyd. 2005. *Fundamentals of metal machining and machine tools*. Vol. 198: CRC Press.

Kryzhanivskyy, V, Volodymyr Bushlya, Oleksandr Gutnichenko, IA Petrusha, and J-E Ståhl. 2015. "Modelling and experimental investigation of cutting temperature when rough turning hardened tool steel with PCBN tools." *Procedia Cirp* 31:489–495.

Kumar, P, and JP Misra. 2020. "Optimization of machining parameters during dry cutting of Ti6Al4V using Taguchi's orthogonal array." In *Emerging trends in mechanical engineering*, 229–243. Springer.

Kumar, Pawan, and Joy Prakash Misra. 2019. "Modelling of tool wear for Ti64 turning operation." *Materials Science Forum*.

Kus, Abdil, Yahya Isik, M Cemal Cakir, Salih Coşkun, and Kadir Özdemir. 2015. "Thermocouple and infrared sensor-based measurement of temperature distribution in metal cutting." *Sensors* 15 (1):1274–1291.

Lin, HM, YS Liao, and CC Wei. 2008. "Wear behavior in turning high hardness alloy steel by CBN tool." *Wear* 264 (7–8):679–684.

Lin, Jehnming, Shinn-Liang Lee, and Cheng-I Weng. 1992. "Estimation of cutting temperature in high speed machining."

Loewen, EG, and MC Shaw. 1954. "On the analysis of cutting-tool temperatures." *Transactions of the American Society of Mechanical Engineers* 76 (2):217–225.

Nedić, Bogdan P, and Milan D Erić. 2014. "Cutting temperature measurement and material machinability." *Thermal Science* 18 (suppl. 1):259–268.

Özel, Tugrul. 2009. "Computational modelling of 3D turning: Influence of edge micro-geometry on forces, stresses, friction and tool wear in PcBN tooling." *Journal of Materials Processing Technology* 209 (11):5167–5177.

Pal, Awadhesh, SK Choudhury, and Satish Chinchanikar. 2014. "Machinability assessment through experimental investigation during hard and soft turning of hardened steel." *Procedia Materials Science* 6:80–91.

Pawan, Kumar, and Joy Prakash Misra. 2018. "A surface roughness predictive model for DSS longitudinal turning operation." *DAAAM International Scientific Book* 25:285–296.

Shihab, Suha K, Zahid A Khan, Aas Mohammad, and Arshad Noor Siddiqueed. 2014. "RSM based study of cutting temperature during hard turning with multi-layer coated carbide insert." *Procedia Materials Science* 6:1233–1242.

Stephenson, David A, and John S Agapiou. 2018. *Metal cutting theory and practice*: CRC Press.

Svetlitza, Alex, Michael Slavenko, Tatiana Blank, Igor Brouk, Sara Stolyarova, and Yael Nemirovsky. 2014. "THz measurements and calibration based on a blackbody source." *IEEE Transactions on Terahertz Science and Technology* 4 (3):347–359.

Thakare, Amol, and Anders Nordgren. 2015. "Experimental study and modeling of steady state temperature distributions in coated cemented carbide tools in turning." *Procedia CIRP* 31:234–239.

Valera, Harsh Y, and Sanket N Bhavsar. 2014. "Experimental investigation of surface roughness and power consumption in turning operation of EN 31 alloy steel." *Procedia Technology* 14:528–534.

Veiga, F, M Arizmendi, A Jiménez, and A Gil Del Val. 2021. "Analytical thermal model of orthogonal cutting process for predicting the temperature of the cutting tool with temperature-dependent thermal conductivity." *International Journal of Mechanical Sciences* 204:106524.

Young, Hong-Tsu. 1996. "Cutting temperature responses to flank wear." *Wear* 201 (1–2):117–120.

Chapter 4

Leaf disease recognition

Comparative analysis of various convolutional neural network algorithms

Vikas Kumar Roy
Ruprecht Karl University of Heidelberg, Heidelberg, Germany

Ganpati Kumar Roy
Sharda University, Greater Noida, India

Vasu Thakur
Ruprecht Karl University of Heidelberg, Heidelberg, Germany

Nikhil Baliyan
Roorkee Institute of Technology, Roorkee, India

Nupur Goyal
Graphic Era Deemed to be University, Dehradun, India

CONTENTS

DOI: 10.1201/9781003434849-4

4.1 INTRODUCTION

Leaf disease is one of the major problems and threats to agriculture and food security [Qiang et al. 2019]. Agriculture tries to fulfill the need of the world's large population, and farmers all over the world spent billions of dollars on disease management. Because of the scarcity of the newest technology support, the disease management control in India is poor. Diseases in plants and agriculture trigger the economy of the country. In developing countries such as India, the major sector contributing 17% to the total GDP is agriculture. Diseases affect the food and crops, causing significant losses to the farmer and threatening food security. It is impossible to manage agriculture without handling the disease problems.

For efficient disease management, the prerequisite step is to discover a disease in the initial phase. To detect disease, a farmer can manually check for any abnormal growth or for the sign of a disease-causing organism in plants, or by observing with naked eyes [Militante et al. 2019]. With the help of the latest technology, the automatic detection of leaf disease from natural images is done by artificial intelligence through deep learning methods. In this study, the researcher detected various leaf diseases in different plants at an early stage.

In this chapter, our objective is to ease disease identification by using various architectures of CNN [Russakovsky et al. 2015]. Firstly, we perform various disease classifications using VGG-16, VGG-19 [Liu et al. 2017], and InceptionV3 [Szegedy et al. 2015, 2016] architecture and achieving promising results. After that, classification is done by DenseNet-121 [Too et al. 2019], which achieved the best result among these three architectures. The comparative analysis is done through an accuracy graph and confusion matrix. The proposed model is the DenseNet-121 [Huang et al. 2017], in which mathematical equations of DensNet-121 are employed. Then, the reason is explained as to how to achieve the best results of this proposed approach.

The chapter is classified under following sections: "Introduction" (Section 4.1), "Literature review" (Section 4.2), "Dataset" (Section 4.3), "Methodology" (Section 4.4), "Results and discussion" (Section 4.5), and "Conclusion" (Section 4.6).

4.2 LITERATURE REVIEW

The advancement in machine learning resulted in focusing on recognizing the disease. The mostly used deep learning model is CNN. Deep learning is an advanced method, which is essentially a neural network with three or more layers that works like a human brain [Benuwa et al. 2016]. LeCun et al. 2015 have explained the deep learning process; it could fetch the patterns easily. Chakravarthy and Sundaresan (2020) explained the CNN architecture and that the classification of Early Blight was done using the architecture of CNN, the ResNet, and Xception, and this resulted in an accuracy of 99.735% and 99.952%. Kaizhou Li et al. (2020) applied the architecture of CNN, that is, Inception-V3 models, to categorize the various ginkgo leaf diseases and obtained the accuracy after preprocessing and training 2,408 original images and 1,322 original images. The accuracy achieved by Inception-V3 model is 92.3% in a laboratory environment and 93.2% in a field environment. Vimal Adit, V et al. made use of an open-source dataset of infected leaves consisting of 76,000 images and trained three specific models, i.e., LeNet, AlexNet, and Inception-V3, to classify 38 different plant infections. A comparative analysis was also performed among the models, which predicted that Inception-V3 had outperformed the other models by gaining an accuracy of 98% [Vimal Adit et al. 2020]. Qiang et al. (2019) had explained the observations of plant leaf disease recognition on the basis of the model Inception-V3.

Rangarajan et al. (2018) proposed AlexNet and VGG-16 to categorize various leaf diseases. The results were analyzed by the required number of images and variations in batch size. They found AlexNet more accurate than VGG-16.

After analyzing the related works, Roy et al. decided to include DenseNet-121 [Pleiss et al. 2017] architecture in conjunction with VGG-16 [Simonyan and Zisserman, 2014], VGG-19, and Inception-V3 for comparative analysis of various deep learning techniques.

4.3 DATASET

Thakur et al. have trained the model on the Kaggle dataset [Hughes and Marcel, 2015], which comprised of 30,543 images consisting of 13 various affected leaves, which can be seen in Figure 4.1. The size of the image was reduced to ($224 \times 224 \times 3$) and normalized through pixel values division by 255. Table 4.1 shows the description of the dataset in two categories: training and testing datasets, consisting of 24,560 and 5,983 images.

4.4 METHODOLOGY

The procedures in this chapter consist of three major steps: data acquisition, data preprocessing, and image classification. The Kaggle dataset comprises different variants of leaves of apple, grape, maize, orange, potato, and tomato. This dataset contains 13 various leaf diseases comprising unaffected images of recognized plants. In the data preprocessing, we resize the images, which are fed into the classification model [Kulkarni 2018]. The process involves training the various architectures of CNN model (DenseNet-121, VGG-16, VGG-19, and Inception-V3) to identify the type of plant leaf,

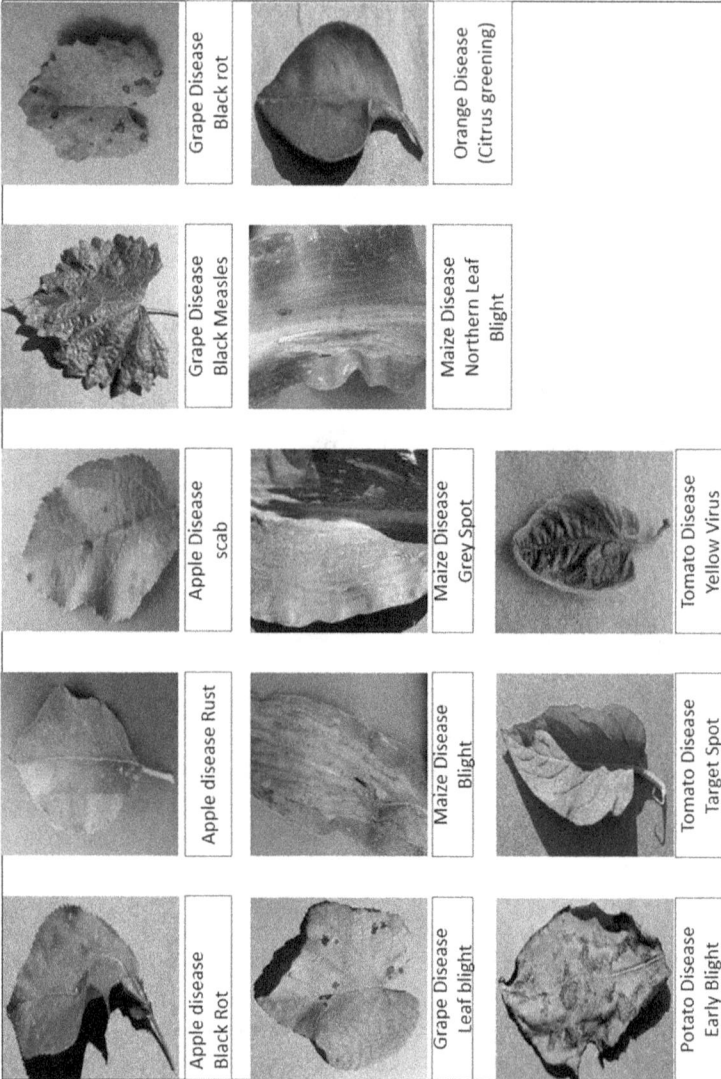

Figure 4.1 Plant disease gathered from the dataset.

Table 4.1 Dataset details

Classes	Number of images
Apple disease: black rot	2,349
Apple disease: rust	2,355
Apple disease: scab	2,349
Grape disease: black measles	2,349
Grape disease: black rot	2,349
Grape disease: leaf blight	2,349
Maize disease: blight	2,349
Maize disease: gray spot	2,349
Maize disease: northern leaf blight	2,349
Orange disease: citrus greening	2,349
Potato disease: early blight	2,349
Tomato disease: target spot	2,349
Tomato disease: yellow virus	2,349
Total	**30,543**

training CNN model [Springenberg et al. 2014] to detect the disease, and validation of model through obtained results.

4.4.1 Convolutional neural network (CNN)

CNN, also known as ConvNet, can discover better when compared with well-connected layer network. It gets high attributes and can identify objects efficiently. CNN is broadly used in various spheres because of its astonishing performance. Additionally, it is beneficial to discover complicated convolutions [Indolia et al. 2018]. It is manifested magnificently for the image classification, translation, and segmentation. We used various CNN architectures like DenseNet-121, VGG-16, VGG-19, and Inception-V3 [Esteva et al. 2017] to extract the images.

4.4.1.1 VGG-19

It is composed of six structures, each of which consists of CNN. It is a model used for preprocessing and utilizing CNN and indeterminate activation of layers [Xiao et al. 2020].

We make use of CNN structure to recognize the critical attributes which are required to classify the leaf disease. To achieve the apparent classification, we utilize various algorithms of machine learning and also compare the acquired accuracy of these algorithms.

Table 4.2 Structure of VGG-16

Layer	Filters of kernel	Size of filter
1 and 2	64	3 by 3
3 and 4	124	3 by 3
5,6 and 7	256	3 by 3
8–13	512	3 by 3

Source: Rezvantalab et al. 2021

4.4.1.2 VGG-16

VGG-16 is a CNN that is used for substantial recognition of images and that accomplishes a precision of 91% in the initial tests in ImageNet, which is the dataset of 15 million images having 22,000 classes [Krizhevsky et al. 2017]. An illustration of the architecture of VGG-16 is described in Table 4.2.

4.4.1.3 Inception V3

Inception-V3 is a CNN that aims on slighter reduced calculative ability. It is implemented to build different improvements and amplifications, specifically breaking of higher-dimensional convolutions [Szegedy et al. 2016], enhanced applications of computation resources [Szegedy et al. 2015], and the usage of a supplemental classifier for the generation of detail labeling. It uses an additional classifier for the transmission of data information of the label to the network at the lower level to make better uniformity in the neural network.

4.4.1.4 DenseNet-121(proposed approach)

DenseNet [Huang et al. 2017] has a compact association compared to other VGG-16, VGG-19, and Inception-V3. It increases the distribution of feature maps and alleviates the required parameters.

The comprehensive structure of the DenseNet can be seen in Figure 4.2. In the proposed architecture, various associations with other CNN models are directly connected from several layers to the successive layers, which can be beneficial for the improvement of the data propagation. Apparently, the formula for the first layer, which acquires the feature maps of previous layers, is as follows:

$$Y^1 = M^1\left(\left[Y^0, Y^1, ..., Y^{l-1}\right]\right)$$

Here, l = layer index

Y^l = result of first layer
$[Y^0, Y^1, ..., Y^{l-1}])$ = feature maps association

M^l = composite function [He et al. 2016] of three successive operations: rectified linear units (RELUs) [Glorot et al. 2011], pooling and convolution [LeCun et al. 1998], and batch normalization [Ioffe and Christian, 2015].

In the DenseNet model, layers facilitate feature maps which are merged through concatenation instead of summation, which is demonstrated in the Figure 4.3.

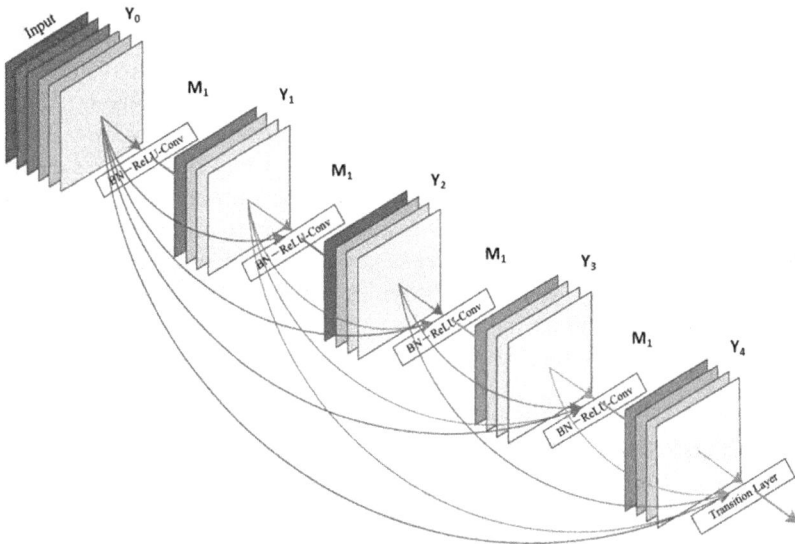

Figure 4.2 DenseNet model.

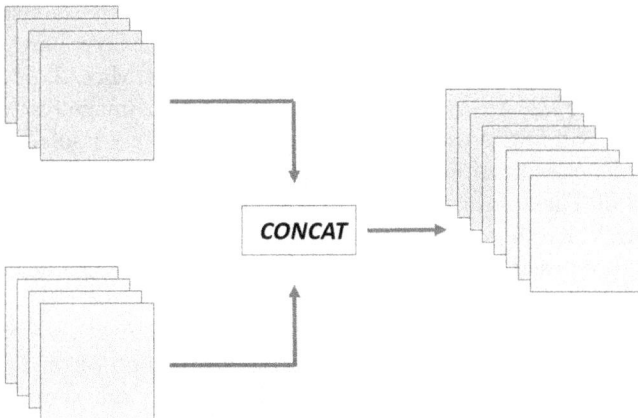

Figure 4.3 Feature maps concatenation.

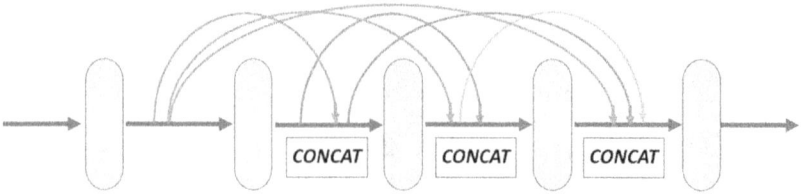

Figure 4.4 Dense connection.

The gradient flow in the dense connections is better than in other models. In Figure 4.4, the gradient flow of the information is shown comprehensively. In Figure 4.4, it is observed that the input signal and loss function is concatenated with each layer in the DenseNet model. DenseNet-121 consists of six dense layers in a block. The result of every layer is directly proportional to the extension of the dense block. DenseNet models provide stability and up-gradation inaccuracy when the number of parameters is increased. DenseNet model needs low computation to attain state-of-the-art accomplishment. Furthermore, DenseNet associates with the features of identity mappings by default. It provides feature reprocesses all over the network. The machine we have implemented in our proposed model constitutes a RAM of 8GB, Intel i5.

4.5 RESULTS AND DISCUSSION

This section illustrates the overall outcome acquired by the proposed DenseNet121 architecture. It is concluded that the accuracy of our approach is 91.75% as compared to other architectures. Comprehensively, we analyzed four scenarios altogether.

Firstly, the VGG-19 structure is implemented over the dataset. It is a model comprising of 19 layers of 16 CNN, where 3 layers are well connected, 5 layers are Max Pool, and 1 layer is of SoftMax. 224*224 RGB images of static size are to be fed as an input. For the image resolution, the dimensional padding is used, and then 2*2 pixel is max-pooled, and afterward, for the classification of model, an RELU is performed for better time computation. The accuracy achieved by VGG-19 architecture is 90.86%.

Secondly, VGG-16 architecture is implemented that accomplished a better precision. It consists of 13 CNN, along with three well-connected layers, and then accomplished a precision of 91.26%.

In scenario 3, an Inception-V3 is implemented that makes use of an additional classifier for the transmission of data information of the label to the network at the lower level for better uniformity in the neural network. Inception-V3 architecture accomplished a precision of 90.98%.

Table 4.3 Leaf disease classification dataset model results

Models	Accuracy (%)
VGG-19	90.86
VGG-16	91.26
Inception-V3	90.98
DenseNet121	91.75

Finally, we proposed the DenseNet method because it executes deep supervision and uses one classifier on the top side of the network. It is less sophisticated as it allows layers to attain feature maps through transition layers also. DenseNet is very efficient when applied to the classification of images; it is effective regarding the number of parameters, has less computation, and has high level of accuracy. It comprises a convolutional layer that extracts low-level image features, and there is a transition layer between consecutive blocks. The parameters in DenseNet are more illustrative on average and reduce the gradient problem. The feature map in this model can be concatenated easily due to the same size. The signals that tend to error can be transmitted easily and directly due to which smooth decision boundaries are given efficiently even when the sample data is not adequate.

To analyze the performance of the DenseNet model, the author compared the DenseNet with other CNNs architectures. Table 4.3 depicts the outcome of various models on the publicly available dataset in leaf disease image classification.

To attain the best analysis, the author observed two categories of the curve: training and validation accuracy vs. epoch curve and training and validation loss vs. epoch curve.

In Figure 4.5, the disparity between the training accuracy and validation accuracy is much higher than any other curve, implying that the curve is neither overfitting nor underfitted. A model has an overfitted curve if the model is trained with even the noise in the data and fetched by the model as a concept, and underfitting disrupts the precision of the model.

In Figure 4.6, both increasing trend of train loss and decreasing trend of validation loss in the DenseNet model imply that the model has a very high learning rate compared to all other models.

4.6 CONCLUSION

The authors have provided an approach to recognize the detection of leaf disease. Our model employs DenseNet to fetch the major features of the disease in image patches of small size. The goal of our work is to find the most appropriate model to identify leaf disease. Observations on the Kaggle

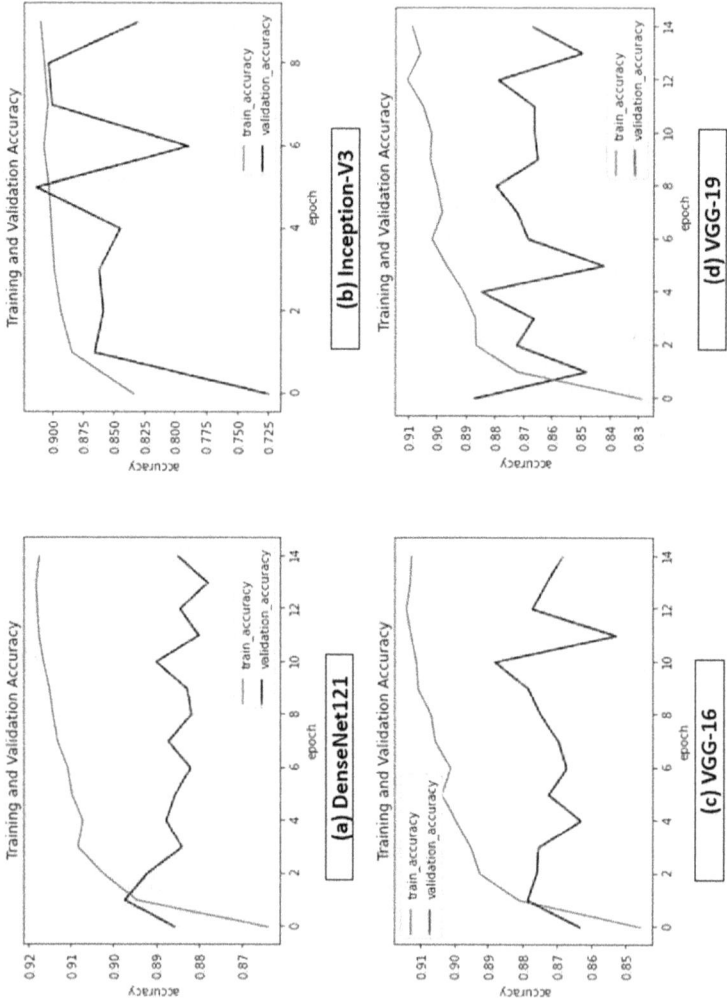

Figure 4.5 Training and validation accuracy of models.

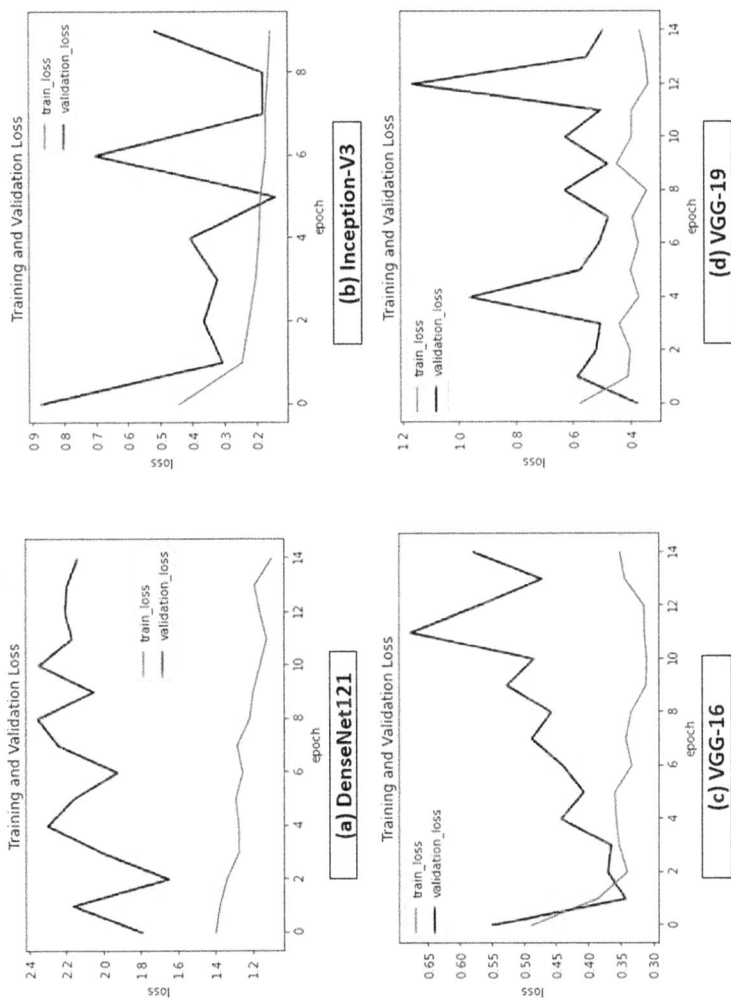

Figure 4.6 Training and validation loss.

dataset concluded that our model attained the highest accuracy compared to other CNN algorithms.

The authors' objective is to make use of hyperparameter settings to upgrade DenseNet so that it'd be beneficial for farmers to acquire the best performance results. It is proposed that our method could be an effective and relevant improvement to the agriculture field.

REFERENCES

Benuwa, Ben Bright, Yong Zhao Zhan, Benjamin Ghansah, Dickson Keddy Wornyo, and Frank Banaseka Kataka. "A review of deep machine learning." *International Journal of Engineering Research in Africa* 24 (2016): 124–136.

Chakravarthy, Anirudh Srinivasan, and Raman Sundaresan. "Early blight identification in tomato leaves using deep learning." In *2020 International conference on contemporary computing and applications (IC3A)*, pp. 154–158. IEEE, 2020.

Esteva, Andre, Brett Kuprel, Roberto A. Novoa, Justin Ko, Susan M. Swetter, Helen M. Blau, and Sebastian Thrun. "Dermatologist-level classification of skin cancer with deep neural networks." *Nature* 542, no. 7639 (2017): 115–118.

Glorot, Xavier, Antoine Bordes, and Yoshua Bengio. "Deep sparse rectifier neural networks." In *Proceedings of the fourteenth international conference on artificial intelligence and statistics*, pp. 315–323. JMLR Workshop and Conference Proceedings, 2011.

He, Kaiming, Xiangyu Zhang, Shaoqing Ren, and Jian Sun. "Identity mappings in deep residual networks." In *European conference on computer vision*, pp. 630–645. Springer, Cham, 2016.

Huang, Gao, Zhuang Liu, Laurens Van Der Maaten, and Kilian Q. Weinberger. "Densely connected convolutional networks." In *Proceedings of the IEEE conference on computer vision and pattern recognition*, pp. 4700–4708. 2017.

Hughes, David, and Salathé Marcel. "An open access repository of images on plant health to enable the development of mobile disease diagnostics." *arXiv preprint arXiv:1511.08060* (2015).

Indolia, Sakshi, Anil Kumar Goswami, Surya Prakesh Mishra, and Pooja Asopa. "Conceptual understanding of convolutional neural network-a deep learning approach." *Procedia Computer Science* 132 (2018): 679–688.

Ioffe, Sergey, and Szegedy Christian. "Batch normalization: Accelerating deep network training by reducing internal covariate shift." In *International conference on machine learning*, pp. 448–456. PMLR, 2015.

Krizhevsky, Alex, Ilya Sutskever, and Geoffrey E. Hinton. "Imagenet classification with deep convolutional neural networks." *Communications of the ACM* 60, no. 6 (2017): 84–90.

Kulkarni, Omkar. "Crop disease detection using deep learning." In *2018 Fourth international conference on computing communication control and automation (ICCUBEA)*, pp. 1–4. IEEE, 2018.

LeCun, Yann, Yoshua Bengio, and Geoffrey Hinton. "Deep learning." *Nature* 521, no. 7553 (2015): 436–444.

LeCun, Yann, Léon Bottou, Yoshua Bengio, and Patrick Haffner. "Gradient-based learning applied to document recognition." *Proceedings of the IEEE* 86, no. 11 (1998): 2278–2324.

Li, Kaizhou, Jianhui Lin, Jinrong Liu, and Yandong Zhao. "Using deep learning for Image-Based different degrees of ginkgo leaf disease classification." *Information* 11, no. 2 (2020): 95.

Liu, Bin, Yun Zhang, Dong Jian He, and Yuxiang Li. "Identification of apple leaf diseases based on deep convolutional neural networks." *Symmetry* 10, no. 1 (2017): 11.

Militante, Sammy V., Bobby D. Gerardo, and Nanette V. Dionisio. "Plant leaf detection and disease recognition using deep learning." In *2019 IEEE Eurasia conference on IOT, communication and engineering (ECICE)*, pp. 579–582. IEEE, 2019.

Pleiss, Geoff, Danlu Chen, Gao Huang, Tongcheng Li, Laurens Van Der Maaten, and Kilian Q. Weinberger. "Memory-efficient implementation of densenets." *arXiv preprint arXiv:1707.06990* (2017).

Qiang, Zhenping, Libo He, and Fei Dai. "Identification of plant leaf diseases based on inception V3 transfer learning and fine-tuning." In *International Conference on Smart City and Informatization*, pp. 118–127. Springer, Singapore, 2019.

Rangarajan, Aravind Krishnaswamy, Raja Purushothaman, and Aniirudh Ramesh. "Tomato crop disease classification using pre-trained deep learning algorithm." *Procedia Computer Science* 133 (2018): 1040–1047.

Rezvantalab, Amirreza, Samir Mitha, and April Khademi. "Alzheimer's disease classification using vision transformers." (2021).

Russakovsky, Olga, Jia Deng, Hao Su, Jonathan Krause, Sanjeev Satheesh, Sean Ma, Zhiheng Huang et al. "Imagenet large scale visual recognition challenge." *International Journal of Computer Vision* 115, no. 3 (2015): 211–252.

Simonyan, Karen, and Andrew Zisserman. "Very deep convolutional networks for large-scale image recognition." *arXiv preprint arXiv:1409.1556* (2014).

Springenberg, Jost Tobias, Alexey Dosovitskiy, Thomas Brox, and Martin Riedmiller. "Striving for simplicity: The all convolutional net." *arXiv preprint arXiv:1412.6806* (2014).

Szegedy, Christian, Wei Liu, Yangqing Jia, Pierre Sermanet, Scott Reed, Dragomir Anguelov, Dumitru Erhan, Vincent Vanhoucke, and Andrew Rabinovich. "Going deeper with convolutions." In *Proceedings of the IEEE conference on computer vision and pattern recognition*, pp. 1–9. 2015.

Szegedy, Christian, Vincent Vanhoucke, Sergey Ioffe, Jon Shlens, and Zbigniew Wojna. "Rethinking the inception architecture for computer vision." In *Proceedings of the IEEE conference on computer vision and pattern recognition*, pp. 2818–2826. 2016.

Too, Edna Chebet, Li Yujian, Sam Njuki, and Liu Yingchun. "A comparative study of fine-tuning deep learning models for plant disease identification." *Computers and Electronics in Agriculture* 161 (2019): 272–279.

Vimal Adit, V., C. V. Rubesh, S. Sanjay Bharathi, G. Santhiya, and R. Anuradha. "A comparison of deep learning algorithms for plant disease classification." In *Advances in cybernetics, cognition, and machine learning for communication technologies*, pp. 153–161. Springer, Singapore, 2020.

Xiao, Jian, Jia Wang, Shaozhong Cao, and Bilong Li. "Application of a novel and improved VGG-19 network in the detection of workers wearing masks." In *Journal of Physics: Conference Series*, vol. 1518, no. 1, p. 012041. IOP Publishing, 2020.

Chapter 5

On the validity of parallel plate assumption for modelling leakage flow past hydraulic piston-cylinder configurations

Rishabh Gupta, Jatin Prakash, Ankur Miglani, and Pavan Kumar Kankar

Indian Institute of Technology, Indore, India

CONTENTS

5.1 INTRODUCTION

The hydraulic actuators used in state-of-the-art hydraulic circuits for both stationary and mobile fluid power applications rely on piston-cylinder arrangements (single or double acting) that convert the pressure of the confined fluid into linear motion and to generate high force. In addition to these functionalities, the hydraulic actuators have been incorporated into a gamut of different configurations, smart designs, and sizes due to their simple geometry. This versatility not only opens the possibility of more-innovative designs but also makes many custom applications a reality, which could not have been realized in practice without the actuators. While on the one end the hydraulic actuators offer several advantages, on the other they are highly prone to leakage due to wear because they operate under very tight clearances: often less than 20 μm. Leakage is the key performance-deteriorating effect of wear (due to contaminants or sliding wear) that increases the energy consumption of the system and impairs its performance. For instance, studies [1–11] indicated that an increase in the

DOI: 10.1201/9781003434849-5

piston-cylinder clearance from 20 μm to 200 μm resulted in a leakage flow of ~57 lpm and a pressure of ~70 bar for a hydraulic cylinder which was originally designed to generate a clamping force of 500 tons with a flow rate of ~170 lpm and pressure of ~207 bar. This indicates that approximately 33% of the flow rate is lost via the leakage, which can significantly deteriorate the system efficiency. Furthermore, a leak in a reciprocating piston-cylinder system can induce uneven flow rate and pressure fluctuations, which may induce fatigue in the actuator as well as the upstream and downstream components. This in turn can lead to their premature failure, thereby affecting the system performance, and increasing the maintenance downtime and the associated costs.

To enable leakage fault detection at an early stage and increase the system predictability and reliability, several previous studies have focused on the leakage fault diagnosis [1–7] of piston-cylinder arrangements in hydraulic actuators, axial piston pumps, and DC valves, to name a few. Tang et al. [3] adopted a model-based approach to simulate the leakage flow past a piston-barrel clearance in an axial piston pump, where the clearance was modelled as that between two parallel plates of infinite length and width. The model predictions of the leakage flow rate were not validated against the experiments. Li et al. [4] developed a model-based approach to predict the performance of an axial piston pump as a function of the increasing severity of leakage fault (*i.e.*, annular clearance h) of 30 μm, 60 μm, and 90 μm. Like Ref. [3], the leakage flow through the annular clearance was modelled as a pressure-driven flow between two fixed parallel plates separated by a constant width. While the model predicted the leakage flow rate to vary as ~ h^3, the experimental data showed that the variation in the pressure and the flow rate at the pump discharge was insignificant with increase in h from 30 μm to 90 μm, and, therefore, did not follow the cubic dependence. Bergada and Watton [7] also developed a leakage flow model to mimic the leakage flow past the piston-barrel annular clearance in a swash plate pump. The expression for leakage flow rate was derived by applying the continuity equation, the 1D-Reynold's lubrication equations, and the Couette-Poiseuille flow equations to the gap between the piston and cylinder, assuming it to be the gap between two parallel plates with grooves. While the effect of leakage fault severity was determined numerically, it was not validated experimentally. A few studies have also explored the effects of different operating parameters that affect the leakage flow such as the working pressure and flow rate, pump rpm, the fluid viscosity, fluid density, laminar versus turbulent flow, etc. [9], and the factors affecting flow geometry such as cylinder diameter, eccentricity, axial wear profile, and the presences of circumferential grooves [8, 10]. Other studies have proposed the system-level modelling using different methods such as the bond graph simulation technique for leakage fault detection and have demonstrated that dynamic response signals can be improved by examining the leakage resistance through simulation performed for various hydraulic components, including actuator [11–16]. Strmčnik and Majdič [17] demonstrated that by tracking the energy losses in the hydraulic system, the leakage fault (increased clearance between the spool and the valve body) can be detected in a DC valve, which was confirmed through a validation between simulations and experiments. Prakash et al. [18, 19] used contemporary deep learning and classification techniques approaches to estimate the health of a cooling circuit. Monitoring the deterioration of hydraulic components such as valves and other sensitive components in hydraulic circuits using contemporary analysis is critical to keeping time limitations in experimental investigations in check. Some of the recent work establishes importance of machine learning techniques for classification problems in various domains such as fault diagnosis of pumps [22, 23], damage classification of grains [24], detecting surface defects in steel plates [25], etc. While the recent studies have made great strides in leakage detection, state-of-the-art models still assume the annular leakage path between a piston and a cylinder as that between two fixed parallel plates of very large length and width that are separated by a constant distance equal to the annular gap. Additionally, these models lack experimental validation at lab scale or against the standard industrial practices for predicting the magnitude of leakage. Since the amount of leakage is primarily governed by the flow cross-section, it is imperative to develop models that account for an actual geometry of the leakage flow passage (i.e., an annular passage) and enable an accurate prediction of the leakage flow rate for early fault diagnosis.

This chapter draws a comparison between two models, namely, the parallel plate model, and the annular flow model, which are used for predicting the leakage flow past the annular clearance between a hydraulic piston and a cylinder arrangement. Subsequently, these models are validated against the orifice plate model, which is used as a standard industrial practice for calculating the leakage magnitude. The comparison with the orifice model is made in terms of the variation in three parameters: the leakage flow rate, the net flow rate, and the volumetric efficiency as a function of the severity of leakage fault (i.e., increasing annular clearance). Subsequently, it is demonstrated that the annular leakage model is better at capturing the parametric trends predicted by the orifice model and, therefore, more suitable of leakage fault diagnosis of piston-cylinder assemblies in hydraulic systems, as shown in Figure 5.1.

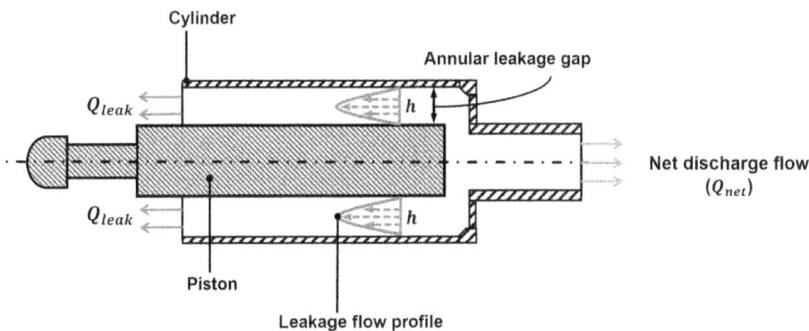

Figure 5.1 Sectional view of the piston-cylinder arrangement showing the oil leakage path through a uniform annulus of width h.

5.2 THE LEAKAGE FLOW MODELS

This section describes the parallel plate and the annular leakage flow models and derives the mathematical expression of leakage flow rate past the annular clearance between a piston and a cylinder, often encountered in high-pressure fluid power systems. The initial assumptions and the key steps in model development are described in detail.

5.2.1 Parallel plate model

In the parallel plat model, the leakage flow path between the piston and the cylinder is assumed as that between two parallel plates of infinite length, as shown in Figure 5.2.

The 2D leakage flow between the two fixed parallel plates separated by a distance h is governed by the pressure difference between the high-pressure discharge and the drain. The expression for this pressure-driven leakage flow rate represents the Poiseuille flow and can be derived using the Navier-Stokes momentum equations along with the continuity equation [4].

Assuming axial symmetry and that the plates are very wide and long such that the flow is only in the axial direction, i.e., $u \neq 0$ but the radial and circumferential velocity components are zero, the continuity equation for the flow is given by:

$$\frac{\partial u}{\partial x} + \frac{\partial v}{\partial r} = 0 \rightarrow \frac{\partial u}{\partial x} = 0 \, \text{or} \, u = u(r) \, \text{only} \tag{5.1}$$

Equation 5.1 indicates that only the streamwise velocity component is significant that varies across the annular gap. The x-momentum Navier-Stokes equation is given by:

$$\left(u \frac{\partial u}{\partial x} + v \frac{\partial u}{\partial x} \right) = -\frac{1}{\rho} \frac{\partial P}{\partial x} + \frac{\mu}{\rho} \left(\frac{\partial^2 u}{\partial x^2} + \frac{\partial^2 u}{\partial r^2} \right) + a_x \tag{5.2}$$

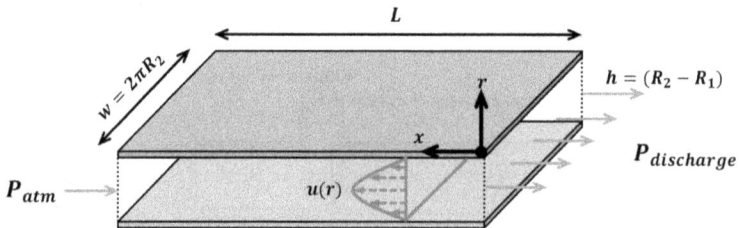

Figure 5.2 Schematic diagram showing pressure-driven leakage flow past the clearance between two fixed parallel plates (parallel plate flow model) of very large length and width.

Substituting $v = 0, \dfrac{\partial u}{\partial x} = 0$ from Eq. 5.1, and the acceleration in x-direction $a_x = 0$, the momentum equation governing leakage flow between parallel plates reduces to:

$$\mu\left(\frac{\partial^2 u}{\partial r^2}\right) = \frac{\partial P}{\partial x} = \frac{\Delta P}{L} < 0 \tag{5.3}$$

Since the radial and circumferential velocity components are zero, the r–momentum, and θ– momentum equations reduce to:

$$\frac{\partial P}{\partial r} = 0, \text{ and } \frac{\partial P}{\partial \theta} = 0 \text{ or } P = P(x) \text{ only} \tag{5.4}$$

Equations 5.3 and 5.4 indicate that the pressure gradient in the axial direction is the total and the only gradient that drives the leakage flow past the annular gap. By integrating Eq. 5.3 twice, noting that $\dfrac{du}{dr} = 0$ at the centreline, and applying the no-slip boundary condition: i.e., $u = 0$ at $r = 0$ and $r = h$, the parabolic Poiseuille velocity profile is obtained as follows:

$$u = \frac{\Delta P}{2\mu L}(h - r)r \tag{5.5}$$

The leakage flow rate Q_{leak} is determined as follows:

$$Q_{leak} = \int_0^h u(r) \cdot w\, dr \tag{5.6}$$

The parallel plate assumption annular leakage flow area between the parallel plates is assumed to be a rectangular cross-section of constant width $w = 2\pi R_2$, and, therefore, the flow area is not a variable parameter while integrating Eq. 5.6. Considering the flow area of constant width, the expression for leakage flow rate is given by:

$$Q_{leak} = \frac{\pi R_2 h^3}{6\mu} \frac{\Delta P}{L} \tag{5.7}$$

Equation 5.7 indicates that the leakage flow rate has a cubic dependence ($Q_{leak} \propto h^3$) on the annular clearance when the flow is that between two parallel plates of very large length and width compared to the clearance.

5.2.2 Annular flow model

In the annular flow model, the leakage flow path is modelled as an annulus of uniform width $h(=R_2 - R_1)$ between two concentric cylinders of radii R_1 and R_2, respectively, which is an appropriate geometric representation of flow geometry. The annular geometry is shown schematically in Figure 5.3. Due to wear, the leakage severity increases with an increase in the annular gap, which may range from $O(10\mu m)$ to $O(100\mu m)$. At such small clearances that are typically encountered in hydraulic systems, the Reynold's number (Re) of the leakage flow is measured to be between 1 and 100, which indicates that the flow is laminar.

The expression for leakage flow rate through a uniform annulus can be derived from the mass conservation and the Navier-Stokes equations for incompressible viscous fluid flow [11]. Neglecting the effects of gravity and assuming axial symmetry of the flow $\left(\frac{\partial}{\partial\theta}=0\right)$, the mass conservation equation for the fully developed (radial and angular components of velocity are zero) laminar flow through the annulus is given in cylindrical coordinates as [12, 13] follows:

$$\frac{\partial u}{\partial x} = 0 \text{ or } u = u(r) \tag{5.8}$$

With the continuity Eq. 5.8, the radial momentum equation reduces to $\frac{\partial P}{\partial r} = 0$, i.e., the pressure drop varies along the leakage path only in the streamwise direction ($P = P(x)$ only), while the x-momentum equation reduces to:

$$\frac{\mu}{r}\frac{d}{dr}\left(r\frac{du}{dr}\right) = \frac{dP}{dx} \tag{5.9}$$

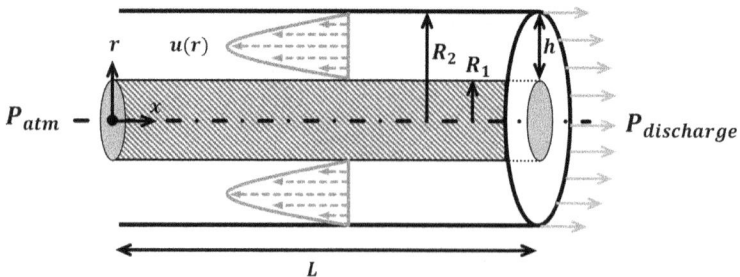

Figure 5.3 Schematic diagram showing pressure-driven leakage flowing past the clearance between two concentric cylinders (annular flow model), showing that the flow area changes with radius.

Here, $\dfrac{dP}{dx} = K$ (constant) < 0. Note that in Eq. 5.9 the convective accelera-

tion term $\rho u \dfrac{\partial u}{\partial x}$ is neglected because from the continuity equation $\dfrac{\partial u}{\partial x} = 0$.

The linear Eq. 5.9 can be integrated twice to obtain the radial variation of the axial velocity as follows:

$$u(r) = \frac{Kr^2}{4\mu} + K_1 \ln(r) + K_2 \tag{5.10}$$

Here, the constants K_1 and K_2 can be obtained using the no-slip boundary conditions at surface of the piston $r = R_1$, and inner surface of the cylinder $r = R_2$, respectively.

$$u(r = R_2) = 0 = \frac{KR_2^2}{4\mu} + K_1 \ln(R_2) + K_2 \tag{5.11}$$

$$u(r = R_1) = 0 = \frac{KR_1^2}{4\mu} + K_1 \ln(R_1) + K_2 \tag{5.12}$$

By solving for K_1 and K_2 from Eqs. 5.11 and 5.12 and plugging into Eq. 5.10, the final velocity profile for the leakage flow through the annular passage is obtained as follows:

$$u(r) = \frac{-K}{4\mu}\left[R_2^2 - r^2 + \frac{R_2^2 - R_1^2}{\ln(R_1/r)} \ln\left(\frac{R_2}{r}\right) \right] \tag{5.13}$$

Further, using Eq. 5.13, the steady-state leakage flow rate can be calculated as follows:

$$Q_{leak} = \int_{R_1}^{R_2} u(r) \cdot 2\pi r\, dr = \frac{-K\pi}{8\mu}\left[R_2^4 - R_1^4 - \frac{\left(R_2^2 - R_1^2\right)^2}{\ln(R_2/R_1)} \right] \tag{5.14}$$

Equation 5.14 can be recast in terms of the magnitude of wear (or the width of the annulus) $h = R_2 - R_1$ as follows:

$$Q_{leak} = \frac{-\pi}{8\mu} h R_1^3 \left(2 + \frac{h}{R_1}\right)\left[1 + \left(1 + \frac{h}{R_1}\right)^2 - \frac{h\left(2 + \dfrac{h}{R_1}\right)}{R_1 \ln\left(1 + \dfrac{h}{R_1}\right)} \right]\frac{\Delta P}{L} \tag{5.15}$$

where the flow resistance

$$F_R = \frac{\Delta P}{Q_{leak}} = \frac{8\mu L}{\pi h R_1^3 \left(2 + \dfrac{h}{R_1}\right)\left[1 + \left(1 + \dfrac{h}{R_1}\right)^2 - \dfrac{h\left(2 + \dfrac{h}{R_1}\right)}{R_1 \ln\left(1 + \dfrac{h}{R_1}\right)}\right]}$$

From Eq. 5.15, it is apparent that the leakage flow rate has a complex dependence on the annular clearance h: $Q_{leak} = f\left(h, \ln\left(1 + \dfrac{h}{R_1}\right)\right)$ as opposed to the direct $Q_{leak} \propto h^3$ dependence obtained from the flat plate model (Eq. 5.7).

5.3 RESULTS AND DISCUSSION

5.3.1 Comparison of leakage flow models

The leakage flow rate derived through both the models (parallel plate and the annular), described in Section 5.2, represents a pressure-driven flow, which is governed by the Navier-Stokes momentum equation in the axial direction. However, the final leakage flow equations obtained from the two models are not the same. This is due to the difference in their initial assumptions and the key steps during the model formulation. These differences are enlisted in Table 5.1 and show a comparison between the models.

In comparing the parallel plate and the annular flow models for predicting the leakage flow, it is apparent that the parallel plate model relies on

Table 5.1 A comparison of parallel plate versus the annular flow model for leakage fault detection in hydraulic piston-cylinder arrangements

Parameter	Parallel plate model	Annular model
Flow geometry	Fixed parallel plates of very large width and length	Uniform annular gap between hollow cylinders of radii R_1 and R_2
Flow area (A_{flow})	$2\pi R_2 \cdot h$ = Constant	$\int_{R_1}^{R_2} 2\pi r \cdot dr$ = Variable
Mean velocity (u_{mean})	$\tfrac{2}{3} u_{max}$	$\tfrac{1}{2} u_{max}$
Leakage flow rate (Q_{leak})	$Q_{leak} \propto h^3$	$Q_{leak} \propto (h, \ln h)$

several restrictive assumptions: First, the flow area is assumed to be constant with the radius. This is restrictive because the annular flow geometry inherently dictates that the flow area varies with the radius. Accordingly, the use of cylindrical coordinates in deriving the leakage flow rate is more justified compared to the Cartesian coordinates used in the parallel plate model. Secondly, the assumption of constant width $w = 2\pi r$ in the parallel plate model means that the model prediction for the leakage flow rate will vary depending on the value of radius r that is chosen as an input, i.e., R_1 or R_2 or $(R_1 + R_2)/2$. Thirdly, for a pressure gradient-driven flow between parallel plates to be essentially axial, i.e., the radial and the circumferential velocity components are zero while the axial component is non-zero, the parallel plates must be very wide and long such that $L >> h$, and $2\pi R_1 >> h$. However, this may not be the case with several piston-cylinder arrangements. For instance, in an axial piston pump, the annular clearance may be ~ 100 µm under the leakage fault conditions, which is of the same order of magnitude or sometimes even comparable (as small as 800 µm [10]) to the length of the grooves on the piston. Finally, the cubic dependence of leakage flow rate on the annular clearance ($Q_{leak} \propto h^3$; Eq. 5.7) dictates that for a fixed pressure drop, 2.2 times increase in the annular clearance would result in an increase in the leakage flow rate by ~10 times (an order of magnitude). However, many previous experimental studies [3–7] have shown that for $h < 100$ µm an increase in the annular clearance by as much as three times has not resulted in any significant change in the leakage flow rate, the discharge pressure, or the discharge flow rate. For example, Bergada et al. [7] experimentally demonstrated that increasing the piston-cylinder clearance from 5 µm to 8 µm (60% increase) in all the cylinders of an axial piston pump resulted in just 6.5% drop in the total flow rate at the pump outlet. In another experimental study, Li [4] demonstrated that increasing the annular clearance from 30 µm to 90 µm (three times) in a single piston of a swash plate pump caused a decrease in the mean leakage flow rate by less than 10%, while the mean pressure at the pump discharge reduced by less than 5%. Based on this discussion, it can be inferred that the concentric cylinder geometry is a more appropriate representation of the leakage path for analysing the leakage flow mechanics between a sliding piston and the cylinder.

5.3.2 Validation against the orifice plate model

While the discussion in the previous section (Section 5.3.1) indicates that the annular flow model is more suitable for predicting the leakage flow rate in an actual scenario, it is crucial to ascertain the utility of the models by calibrating and validating them against the standard modelling practices that are adopted by the industry personnel. One extensively used modelling approach, which is used for a first-hand estimate of the magnitude of leakage, is flow past an orifice with a throat area equivalent to that of the

annular leakage gap [20]. This same approach has also been adopted experimentally for lab-scale testing by both the industry [21] and the industry-
academia joint centres [20, 21] for simulating artificial leakage in axial
piston pumps; where a fixed amount pump discharge flow is by-passed
back to the case drain via a pressure control servo valve or an orifice plate
(a high-impedance device). Therefore, in this section, the leakage flow past
the piston-cylinder configuration in an axial piston pump is compared
using the parallel plate and the annular flow model and calibrated against
the orifice-based leakage flow model, in terms of the variation of the leakage flow rate, the net flow at the discharge, and the pump's volumetric
efficiency as a function of the leakage fault severity (i.e., increasing annular
clearance). To this end, the model that offers a closer fit to the trends predicted by orifice-based leakage model can be considered to be more suitable
for predicting the leakage flow rate.

Figure 5.4 shows an open-loop flow network architecture of a piston-
cylinder configuration of an axial piston pump that is developed using
Simscape in the Simulink environment of MATLAB® 2020b. The leakage
flow via the annular gap between the piston and cylinder is modelled as
a laminar flow through an orifice, whose throat area can be changed
to account for a varying degree of leakage fault ranging from 0 to 100
microns.

To simulate the pressure-driven leakage flow, the orifice is placed between
two reservoirs maintained at different pressures: one at mean discharge
pressure of ~104.6 bar (high-pressure reservoir) and the other at atmo-
spheric pressure of ~1.01 bar. The model consists of an input flow line from
high-pressure reservoir to the actuator and of a leakage flow line from the
orifice to the low-pressure reservoir. Inline flow sensors are placed in both
these lines to collect the input and the leakage flow rate and display this data
on the respective panels. The leakage flow equation based on the orifice
model is given by:

$$\left(Q_{leak,\,orifice}\right)_{th} = C_d \cdot A_{d_i} \cdot \sqrt{\frac{2\left|P_{pc,i} - P_{atm}\right|}{\rho}} \cdot \mathrm{sgn}\left(P_{pc,i} - P_{atm}\right) \tag{5.16}$$

Here, i is the index (i = 1, 2, ... n : number of pistons in an axial piston
pump), $P_{pc,\,i}$ is the pressure inside i^{th} cylinder, and A_{d_i} is the discharge area of
the orifice. As shown in Figure 5.5, A_{d_i} depends on the angular position of
the piston in an axial piston pump and, therefore, varies temporally due to
the rotational motion of the pump barrel, i.e., $A_{d_i} = f(\theta, t)$, where $\theta = \omega t$, and
ω is the pump's angular velocity given by:

$$\omega = \frac{d\theta}{dt} = 2\pi N = \text{constant} \tag{5.17}$$

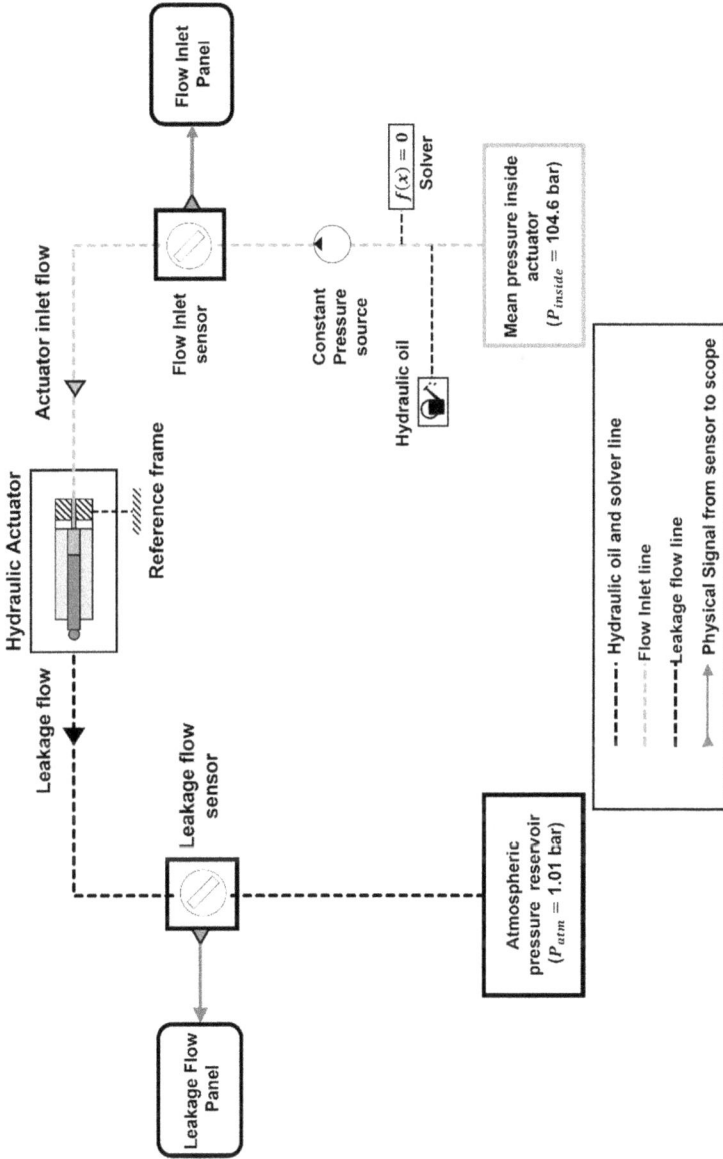

Figure 5.4 An open-loop flow network model developed in MATLAB® Simscape for simulating leakage fault in a hydraulic actuator.

On integrating Eq. 5.17, a linear relationship is obtained between the angular displacement (θ) and time (t):

$$\theta = \omega t + \text{constant} \tag{5.18}$$

Since, the angular displacement and hence the area varies linearly with time, starting from $A_{di} = 0$ at $t = t_1$ and ending at $A_{di} = A_{max}$ and $t = t_2$. The linear variation in the open area of the orifice with time is shown schematically in Figure 5.5 (b). The mean area of the orifice from time $t = t_1$ to $t = t_2$ as it translates the port plate is given by:

$$A_{mean} = \left(A_{max} + A_{min}\right)/2 = \left(A_{max} + 0\right)/2 = \left(A_{max}\right)/2 \tag{5.19}$$

From the Equation (5.16), the expression for the mean flow through the orifice is given by:

$$\left(Q_{orifice}\right)_{model} = C_d \cdot A_{mean} \cdot \sqrt{\frac{2\Delta P}{\rho}} = \left(Q_{leak,orifice}\right)_{th}/2 \tag{5.20}$$

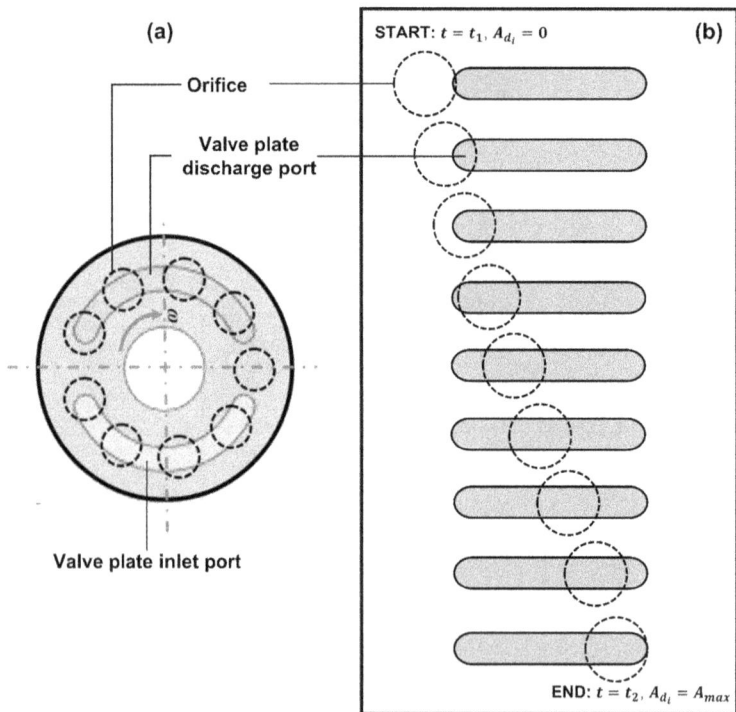

Figure 5.5 (a) Cross-sectional view of the port plate arrangement showing nine equal orifices along the circumference, each with equal area. (b) Schematic showing the gradual variation in the orifice area with time beginning from its opening until its final closure.

Here, ΔP is the differential pressure across the orifice, ρ is the fluid density, and C_d is the coefficient of discharge. When simulating the leakage flow past the piston-cylinder assembly in an axial piston pump by integrating the parallel plate model, and the annular flow model with the flow network (see Figure 5.4), the corresponding discharge areas are calculated as follows:

$$A_{d_i} = A_{max} = \begin{cases} wh & ;t = t_2 \\ \pi\left(R_2^2 - R_1^2\right) & ;t = t_2 \end{cases} \qquad (5.21)$$

The simulations are performed in the MATLAB® R2020b-based modelling environment Simulink 10.2 using the backward-Euler implicit method with a fixed time step of 0.0001 s. For each level of leakage fault, the simulation is run for 1s, which corresponds to several cycles of an axial piston pump, such that the time-mean value of leakage flow can be calculated accurately. The different parameters and operating conditions that are used as inputs to the model are detailed in Table 5.2. Note that these parameters are selected directly based on the specifications of the industrial axial piston pumps [14–16]. The simulation results from this model are presented in terms of the trends in the time-averaged value of three metrics, namely the leakage flow rate, the total flow rate at the pump outlet, and the pump's volumetric efficiency. The variations of these metrics as a function of the severity of the leakage fault are illustrated in Figure 5.6.

Figure 5.6 shows the validation of the parallel plate model and the annular flow model against the orifice model. It is apparent from Figure 5.6a that the orifice model predicts a linear variation in the metrics with increasing severity of leakage fault, while the other two models exhibit a non-linear variation, which can be divided into three regimes. Regime (I): $0 < h < 30$ μm, Regime (II): 30 μm $\leq h \leq$ 50 μm, and Regime (III) $h > 50$ μm. Additionally, the predictions of all three models exhibit the same trends for the three metrics, i.e., increase in the leakage flow rate and a decrease in the net discharge flow and the volumetric efficiency with an increase in the leakage severity. Therefore, for brevity, further discussion is focused on just one metric, the pump's volumetric efficiency, which is the single most important metric of interest from the point of view of ascertaining the pump's condition and the

Table 5.2 Model input parameters for the simulation [14–16]

Orifice model parameter	Value
Pressure difference (ΔP)	10.46 MPa
Density of hydraulic fluid (ρ)	960 Kg/m³
Discharge coefficient (C_d)	0.61
Radius of the piston (R_1)	0.008 m

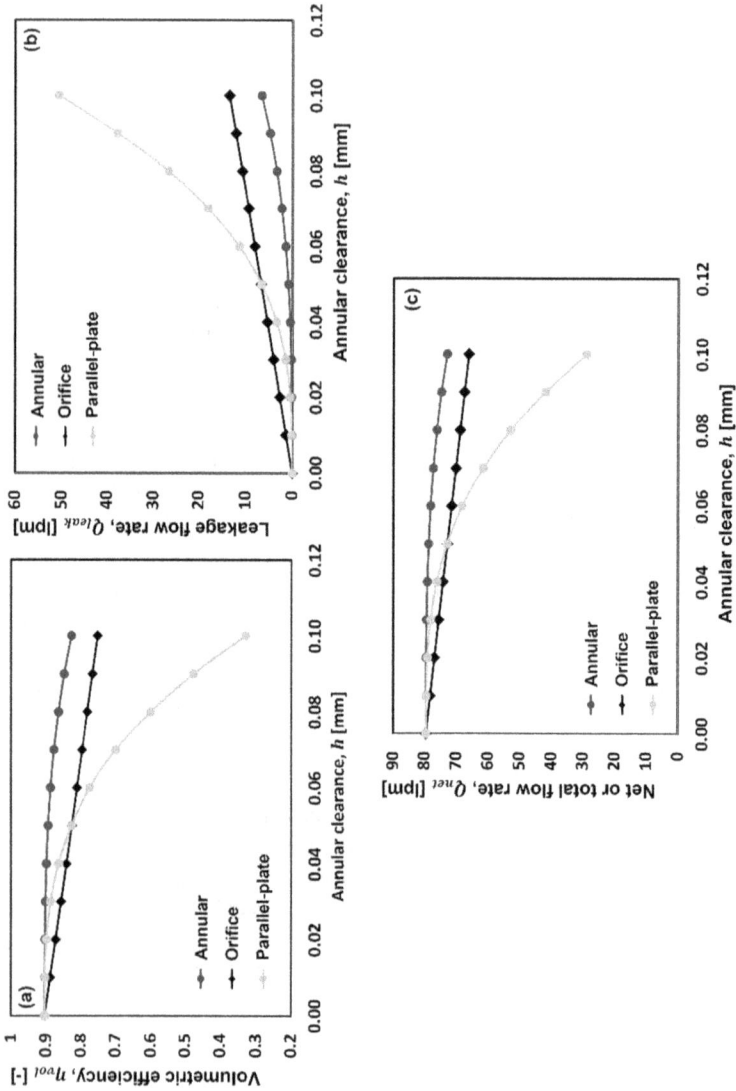

Figure 5.6 Validation of the parallel plate and the annular flow model predictions with the orifice model results based on the variation flow metrics as a function of increasing severity of leakage fault: (a) volumetric efficiency, (b) leakage flow rate, and (c) net discharge flow rate.

need for its maintenance. The volumetric efficiency is determined as the actual flow delivered by the pump for a constant discharge pressure divided by the theoretical flow, at a fixed rpm. In regime (I), where the annular clearance is nominally low (representative of the healthy pump condition [4]), both the parallel plate model and the annular model predict the volumetric efficiency to be 90%, which deviates by less than 2% from the predictions of the orifice model. Further increase in the annular clearance (30 μm $\leq h \leq$ 50 μm) corresponds to regime (II), which marks the occurrence of leakage fault but with a moderate severity [4]. In this regime, the predictions of the parallel plate model and the annular clearance model match closely with the results of the orifice model. The maximum absolute error is 3% (at 30 μm) and 7% (at 50 μm) for the predictions of the parallel plate model and the annular flow model, respectively. Finally, in regime (III), which represents the severe leakage fault condition [4], the predictions begin to deviate from the linear curve of the orifice model. This deviation is significant for the parallel plate model; the maximum absolute error (MAE) increases from nil at 50 μm to as high as 42% at 100 μm. In contrast, the MAE for the annular model remains constant at ~7% in the entire regime (III). Based on this discussion, it can be inferred that the annular flow model shows an excellent agreement with the orifice model, not only because it closely matches the trend predicted by the orifice model for the range of leakage fault (0 μm $\leq h \leq$ 100 μm) but also because the maximum percentage deviation from the orifice model results over this entire range is within 7%. Therefore, the annular leakage model can be considered a better model for predicting the leakage flow compared to the parallel plate model.

5.4 CONCLUDING REMARKS

Leakages in hydraulic systems left unaddressed can cause the company's profit to drip away. This is due to not just the cost of the hydraulic oil itself but also the need for increased housekeeping, increased energy consumption of the equipment, deterioration in the system performance, and an increase in the maintenance costs. To this end, a significant volume of research has been directed towards the modelling-based leakage fault diagnosis of hydraulic piston-cylinder arrangements. Despite these extensive modelling efforts, the leakage flow through the annular clearance between the piston and cylinder is still simulated by assuming the leakage flow to be the pressure-driven Poiseuille flow between two parallel plates of very large length and width, which is not representative of the actual flow geometry, i.e., uniform annulus. Further, the results indicate that the cubic dependence of leakage flow rate on the annular clearance predicted by the parallel plate model is highly restrictive, because this amounts to three orders of magnitude increase in the leakage flow rate with just a single order of magnitude increase the annular clearance (say from 1 to 10 μm), which is seldom observed in practice.

In this chapter, an alternative model involving the annular flow geometry is developed, which overcomes the restrictive assumptions of the parallel plate model. The leakage flow predictions from the new annular model are compared and validated against the standard industry practice of orifice plate-based leakage flow measurement. It is demonstrated that the predictions of the annular flow model closely follow the trends of the leakage flow and volumetric efficiency as a function of the severity of leakage fault (or increasing annular clearance h), and that the maximum deviation from the orifice plate model over the entire range (0 μm $\leq h \leq$ 100 μm) of leakage fault severity is less than 7%. Thus, this chapter demonstrates that the proposed annular leakage model is better at predicting the magnitude of leakage flow (or leakage fault detection) compared to the parallel plate model.

ACKNOWLEDGEMENT

The author would like to thank Science and Engineering Research Board (SERB), DST, India for providing financial support to the project under core research grant (File number: CRG/2022/007409). The authors wish to acknowledge the funding agency.

REFERENCES

[1] Lasaar, R. and M. Ivantysynova. 2001. Gap geometry variations in displacement machines and their effect on the energy dissipation. *International Conference on Fluid Power Transmission and Control* pp. 296–301 Hangzhou, China.

[2] Ivantysynova, M., and R. Lasaar. 2004. An investigation into micro- and macrogeometric design of piston/cylinder assembly of swash plate machines. *International Journal of Fluid Power* 5: 23–26.

[3] Tang, H.S., W. Yang, and Z. Wang. 2019. A model-based method for leakage detection of piston pump under variable load condition. *IEEE Access* 7: 99771–81. https://doi.org/10.1109/ACCESS.2019.2930816.

[4] Li, Z. 2005. Condition monitoring of axial piston pump. PhD diss., University of Saskatchewan. https://harvest.usask.ca/handle/10388/etd-11252005-202705 (accessed 23 May 2021).

[5] Kumar N., B.K. Sarkar, and S. Maity. 2019. Leakage based condition monitoring and pressure control of the swashplate axial piston pump. *ASME 2019 Gas Turbine India Conference, GTINDIA 2019*, vol. 2, American Society of Mechanical Engineers (ASME). https://doi.org/10.1115/GTINDIA2019-2385.

[6] Zhang C., S. Huang, J. Du, X. Wang, S. Wang, and H. Zhang. 2016. A new dynamic seven-stage model for thickness prediction of the film between valve plate and cylinder block in axial piston pumps. *Advances in Mechanical Engineering* 8: 1–15, https://doi.org/10.1177/1687814016671446.

[7] Bergada J. M., S. Kumar, D. L. Davies, and J. Watton. 2012. A complete analysis of axial piston pump leakage and output flow ripples. *Applied Mathematical Modelling* 36: 1731–51. https://doi.org/10.1016/j.apm.2011.09.016.

[8] Wang S., J. Xiang, Y. Zhong, and H. Tang. 2018. A data indicator-based deep belief networks to detect multiple faults in axial piston pumps. *Mechanical Systems and Signal Processing* 112: 154–70. https://doi.org/10.1016/j.ymssp.2018.04.038.

[9] Yoder, V. A., R. Singh, J. T. Dreyer, B. Lilly. 2015. Development of a vibration-based health monitoring procedure using a virtual axial piston pump. Undergraduate Honors Thesis, The Ohio State University. http://hdl.handle.net/1811/86176 (accessed 23 May 2021).

[10] Kumar S., and J. M. Bergada. 2013. The effect of piston grooves performance in an axial piston pump via CFD analysis. *International Journal of Mechanical Sciences* 66: 168–79. https://doi.org/10.1016/j.ijmecsci.2012.11.005.

[11] Roccatello, A., S. Mancò, and N. Nervegna. 2006. Modelling a variable displacement axial piston pump in a multibody simulation environment. *Journal of Dynamic Systems, Measurement, and Control* 129 (4): 456–468. https://doi.org/10.1115/1.2745851.

[12] Kundu, P. K., I. M. Cohen, and D. R. Dowling. 2012. *Fluid mechanics.* 5th ed. Academic Press. https://www.sciencedirect.com/book/9780123821003/fluid-mechanics. [accessed 26 May 2021].

[13] White, F. M. 1986. *Fluid mechanics.* 4th ed. McGraw-Hill. http://vlib.kmu.ac.ir/kmu/handle/kmu/80712. [accessed 23 May 2021].

[14] PVM piston pumps service manual, http://www.eaton.in/ecm/groups/public/@pub/@eaton/@hyd/documents/content/pct_477093.pdf. [accessed 26 July 2021].

[15] Vickers industrial hydraulics manual. 2021. https://fdocuments.in/document/289203703-vickers-industrial-hydraulics-manual.html. [accessed 26 July 2021].

[16] Hydraulic axial piston Eaton Vickers PVB pump. http://www.hydpump.com/pdf/Vickers%20PVB%20MVB%20piston%20pump.pdf. [accessed 26 July 2021].

[17] Strmčnik, E, and F. Majdič. 2017. Comparison of leakage level in water and oil hydraulics. *Advances in Mechanical Engineering* 9 (11). https://doi.org/10.1177/1687814017737723.

[18] Prakash, J., and P. K. Kankar. 2020. Health prediction of hydraulic cooling circuit using deep neural network with ensemble feature ranking technique. *Measurement* 151: 107225. https://doi.org/10.1016/j.measurement.2019.107225.

[19] Prakash, J., P. K. Kankar, and A. Miglani. 2021. Monitoring the degradation in the switching behavior of a hydraulic valve using recurrence quantification analysis and fractal dimensions. *Journal of Computing and Information Science in Engineering* 21.6: 061010. https://doi.org/10.1115/1.4050821.

[20] Bohman, E. 2017. Understanding Buckling Strength of Hydraulic Cylinders. https://www.powermotiontech.com/technologies/cylinders-actuators/article/21887243/understanding-buckling-strength-of-hydraulic-cylinders. [accessed 13 July 2021].

[21] Cavera, S. 2018. Cylinders: The basics and more. https://www.powermotiontech.com/technologies/cylinders-actuators/article/21887822/cylinders-the-basics-and-more. [accessed 21 July 2021].

[22] Ranawat N. S., P. K. Kankar, and A. Miglani. 2020. Fault diagnosis in centrifugal pump using support vector machine and artificial neural network.

Journal of Engineering Research- EMSME Special Issue: 99–111. https://doi.org/10.36909/jer.EMSME.13881.

[23] Ranawat, N. S., A. Miglani, and P. K. Kankar. 2022. Performance of centrifugal pump over a range of composite wear ring clearance. *Journal of the Brazilian Society of Mechanical Sciences and Engineering* 44, 524. https://doi.org/10.1007/s40430-022-03835-x.

[24] Bhupendra, K. Moses, A. Miglani, and P. K. Kankar. 2022. Deep CNN based damage classification of milled rice grains using a high-magnification image dataset. *Computers and Electronics in Agriculture* 195: 106811. https://doi.org/10.1016/j.compag.2022.106811.

[25] Patil, S., A. Miglani, P. K. Kankar, and D. Roy. 2022. Deep learning-based methods for detecting surface defects in steel plates. In *Smart Electrical and Mechanical Systems*, 87–107. Academic Press. https://doi.org/10.1016/b978-0-323-90789-7.00001-4.

Chapter 6

Development of a hybrid MGWO-optimized support vector machine approach for tool wear estimation

Navin Rajpurohit, Jeetesh Sharma, and Murari Lal Mittal
Malaviya National Institute of Technology, Jaipur, India

CONTENTS

6.1 INTRODUCTION

Machine tools are an essential part of modern society, so it is required that machines should be reliable, safe, intelligent, and economical due to rapid changes in living standards and industrialization. However, many accidents occur due to problems like system parts blockage, sudden tool fractures, etc. Proper maintenance can provide a solution to the above issues. There are mainly three types of maintenance: corrective, preventive, and predictive. Corrective maintenance takes place when any component of the machine ultimately fails. Preventive maintenance takes place after a scheduled interval of time. Predictive maintenance takes place based on the component's real-time condition and predicts a future maintenance plan. Predictive maintenance has some advantages over preventive maintenance, like low operation costs.

Maintenance records can be found dating back to ancient Egypt. From 600 BC, an ancient Egyptian document has been found that contains data on the supply of cedar wood required for the maintenance of Amun Ra's sacred boat (Brugsch, 1881). Over the years, the process of maintenance has evolved. It has evolved from reactive (corrective) maintenance to predictive maintenance with trade-offs among quality, time, and cost. The primary aim of modern maintenance management is to reduce both unscheduled and scheduled downtime with optimization of cost, safety, and environmental risk. Production quality, performance, and available time are the primary key performance indicators (KPIs) that affect overall equipment effectiveness (OEE) (Kol, 2007). Tool condition monitoring is a form of predictive maintenance. Predictive maintenance (also known as PdM 4.0) is today's most advanced form of maintenance. It avoids asset (tool) failure by analyzing production data to detect patterns and forecast problems at an early stage. To generate insights and detect patterns and anomalies, integration of big data analytics and artificial intelligence is required. For at least one specific asset (tool), it needs complete real-time tool monitoring in conjunction with useful external information (e.g., environmental data, usage), which alerts based on predictive techniques like regression analysis (Orosz et al., 2015).

It's worth noting that a high-quality product usually entails a high-quality surface finish and good dimensional accuracy. A sharp tool should be kept on hand at all times. A dull tool deforms the surface of the workpiece to a greater extent and may tear, which results in lowering the fatigue resistance. The friction produced by a worn tool generates more heat and raises the cutting temperature, which changes the mechanical properties of the workpiece material (Scay, 1999). Therefore, it is necessary to control tool wear. This chapter describes a new hybrid regression model to predict tool flank wear and identify the most influential variables (input features) that give the best-operating conditions for milling machines.

Indeed, this study aims to determine the tool flank wear (output variable) as a function of ten input variables from 4 to 13 (Table 6.1). Thus, this research aims to examine the performance of support vector machines (SVMs) in conjunction with the modified grey wolf optimization (GWO) technique to predict tool flank wear of milling.

Statistical learning theory and risk minimization functions are the foundation of the SVM algorithm. SVMs are categorized as supervised machine learning used for classification and regression due to their ability to approximate multiple variable functions (Kecman, 2005). To improve the regression accuracy of the SVM model, the modified grey wolf optimization (MGWO) is used to find optimal hyper-parameters of the SVM. GWO is a population-based meta-heuristic. It has been extensively customized to solve a wide range of optimization problems due to its unique characteristics than other SI algorithms, such as very few initial parameters, no initial derivative information required, etc. It is adaptable, simple, scalable, and easy to use, with the ability to reach the appropriate balance between exploration and exploitation in search space, resulting in good solution convergence. As a result, the GWO has recently attracted large research followers from various domains in a short period (Faris et al., 2018). Similar to other nature-inspired meta-heuristics, GWO mimics the hunting nature and social

Table 6.1 Input variables and their description

S.N.	Variable name	Description
1.	Case	Case number (1–16)
2.	Run	Counter for experimental runs in each case
3.	VB (mm)	Flank wear, measured after runs; measurements for VB were not taken after each run
4.	Time (mm)	Duration of the experiment (restarts for each case)
5.	DOC (mm)	Depth of cut (does not vary for each case)
6.	Feed (mm/rev)	Feed (does not vary for each case)
7.	Material	Steel or cast iron
8.	smcAC	AC spindle motor current
9.	smcDC	DC spindle motor current
10.	vib_table	Table vibration
11.	vib_spindle	Spindle vibration
12.	AE_table	Acoustic emission at table
13.	AE_spindle	Acoustic emission at the spindle

hierarchy of grey wolves. Similar to artificial bee colony (ABC) optimization (Nieto et al., 2016), particle swarm optimization (PSO) (Robleda et al., 2015), and ant colony optimization (Panigrahi et al., 2011), GWO also solves the problem by improving particle solution in each iteration. Therefore, a hybrid model of MGWO-optimized support vector machine (MGWO-SVM) was used to predict the milling tool flank wear of an industrial milling process as a function of input operating variables (Figure 6.1).

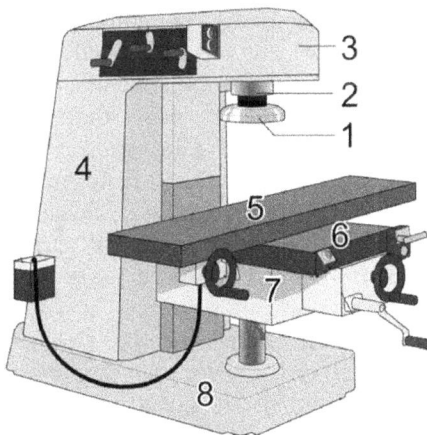

Figure 6.1 Vertical milling machine: (1) Milling tool, (2) Spindle, (3) Top slide or over-arm, (4) Column, (5) Table, (6) Y-axis slide, (7) Knee, (8) Base (Nieto et al., 2016).

According to the review of a SVM, the SVM method is a valuable model for predicting data accurately and has been used in a variety of operational fields successfully, such as rainfall-runoff modeling (Dibike et al., 2001), forest modeling (Nieto et al., 2012), Streamflow forecasting (Jian et al., 2006), a study of water properties (Nieto, 2013), sediment yield forecasting (Mesut, 2008), solar radiation estimation (Chen et al., 2013), evaporation forecasting (Moghaddamnia et al., 2008), lake and reservoir water-level prediction (Khan and Coulibaly, 2006), prediction of the air quality (Nieto, 2013), and so on.

The remaining part of this chapter is described as follows: Section 6.2 consists of the experimental dataset and methods used in this research. Section 6.3 will present results and a discussion of the hybrid model MGWO-SVM. Finally, Section 6.4 draws essential conclusions from this research.

6.2 MATERIALS AND METHODS

6.2.1 Experimental dataset

Experiments on a milling machine under various operating conditions are depicted by the data in this dataset. Tool wear in regular cuts and entry and exit cuts was investigated in particular (Goebel, 1996). The data were collected at multiple locations using three different types of sensors (current sensor, acoustic emission sensor, and vibration sensor). The total number of samples processed in this study was 167, and the factors used are listed in Table 6.1.

There are 16 different cases, each with a different number of runs. The number of runs was determined by the degree of flank wear, computed at irregular intervals between runs until a wear level was reached and sometimes beyond (Goebel, 1996). Flank wear was not continually assessed, and entries were not made when no assessment was conducted. Table 6.1 contains a list of the 16 cases. Figure 6.5 below shows the experimental setup. The Matsuura machining center (MC) spindle and table are powered at 510 V. The machining center's table and spindle each has an acoustic emission sensor and a vibration sensor mounted to them. Before actually entering the computer for data acquisition, each signal of all sensors is amplified as well as filtered before being fed through two roots mean square (RMS) converters. A computer receives the signal from a spindle motor current sensor without further filtering. The criteria for the experiments were chosen based on their industrial suitability and suggested manufacturer's settings (Goebel, 1996).

As a result, the cutting speed was set to 200 meters per minute or 826 revolutions per minute. The depths of the cut were chosen to be 1.5 mm and 0.75 mm, respectively. Two feeds value, 0.5 and 0.25 mm/rev, were used in this study, approximately equal to 413 and 206.5 mm/min, respectively. In addition, two kinds of materials were used (cast iron and stainless steel J45) (Table 6.2).

Table 6.2 Operational conditions of the milling machine

Case	Depth of cut (mm)	Feed (mm/rev)	Material
1.	1.5	0.5	Cast iron
2.	0.75	0.5	Cast iron
3.	0.75	0.25	Cast iron
4.	1.5	0.25	Cast iron
5.	1.5	0.5	Steel
6.	1.5	0.25	Steel
7.	0.75	0.25	Steel
8.	0.75	0.5	Steel
9.	1.5	0.5	Cast iron
10.	1.5	0.25	Cast iron
11.	0.75	0.25	Cast iron
12.	0.75	0.5	Cast iron
13.	0.75	0.25	Steel
14.	0.75	0.5	Steel
15.	1.5	0.25	Steel
16.	1.5	0.5	Steel

6.2.1.1 Data acquisition and processing

The data was sent through a high-speed data acquisition board with a maximum sampling rate of 100 kHz. The signal processing software used the sampled output of the data. For this project, LabVIEW® was used. This software is a graphical language-based general-purpose programming development system (G). An RMS device was used to feed acoustic emission and vibration signals. The use of it smoothens the signal and makes it more processable (Goebel, 1996).

The RMS is proportional to the signal's energy content. For a period of time ΔT, the RMS of a function $f(t)$ is defined by (Goebel, 1996):

$$\text{RMS} = \sqrt{\frac{1}{\Delta T} \int_0^{\Delta T} f^2(t).dt} \qquad (6.1)$$

Here, ΔT = time constant and $f(t)$ = signal function (Figure 6.2).

6.2.1.2 Tool wear

Tool wear occurs in a variety of shapes and sizes. In addition to the cutting edge's normal rounding, crater wear on the rake face is caused by abrasion from the chip sliding across the rake face, and the flank wears out

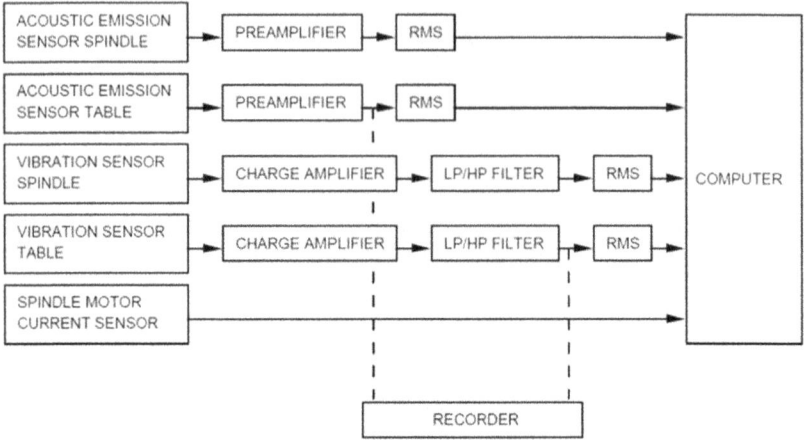

Figure 6.2 NASA Ames milling experimental setup.

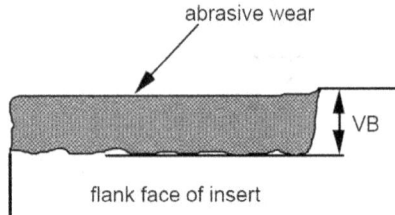

Figure 6.3 Flank wear VB (mm) on the flank face of the milling tool.

due to friction with the workpiece. According to modified Taylor's tool life equation, cutting speed damages more tool life than other parameters (i.e., depth of cut (DOC) and feed), respectively. We used the flank wear VB as a commonly accepted parameter for evaluating tool wear in our experiments (Goebel, 1996) (Figure 6.3).

The distance between the cutting edge and the end of the abrasive wear on the tool's flank face is measured by the flank wear VB. During the experiments, flank wear was identified. The insert was removed from the tool, and the wear was assessed using a microscope.

6.2.2 Support vector regression

SVMs are a collection of methods used for classification and regression tasks that falls under the supervised machine learning category. SVMs were first developed to solve classification problems before being expanded to solve regression problems. Support vector regression (SVR) is the name of the last technique. SVR's model is only based on a subset of the training data because the model's cost function tries to avoid any training data point close to the model prediction (within a threshold ε). The kernel trick is used in SVM if there is non-linearly separable data. When the data in the input space X is not separated, the kernel trick makes things easier. As a result, the kernel maps all data points by transforming the feature space X from low to high dimensions $X \rightarrow \varphi(x)$ (Cristianini and Shawe-Taylor, 2003).

Here's a quick rundown of what SVR is all about. Rather than trying to categorize new, unknown variables x into either of two categories (y = ±1), we would like to estimate a real-valued output y for the observed value t, so we'll use a collection of L data points in the form of $\{x_i|, t_i\}$ our training data, where, $i = 1, 2... L, y \, \varepsilon \, \Re$, and $x \, \varepsilon \, \Re^D$ (John Shawe-Taylor, 2004):

$$y_i = f(x_i) = w.x_i + b \tag{6.2}$$

Here,

y_i = predicted real-valued output
W = normal vector to maximum margin hyperplane
X = feature space vector of D- dimension
"." is the dot product of vectors w and X, which signifies how much the vectors are in the same direction.
b = bias term, which is the distance to the origin of the hyperplane solution.

SVM uses complicated penalty function. The penalty is not imposed if the difference between the predicted value (y_i) and observed value (t_i) is less than the error tolerance (ε), i.e., $|t_i - y_i| < \varepsilon$ the area bounded by $y_i \pm \varepsilon$ is known as ε- an insensitive tube. Another change to the penalty function is that data points are assigned one of the two slack variable penalties outside the tube. If the point is the upper side of the tube (ζ^+) or the lower side of the tube (ζ^-), where, $\zeta^+, \zeta^- \geq 0$ and is defined as follows:

$$t_i - y_i \leq \varepsilon + \zeta^+ \tag{6.3}$$

$$y_i - t_i \leq \varepsilon + \zeta^- \tag{6.4}$$

The next objective is to establish a functional form f that can accurately predict new instances that the SVM has never seen before. This can be obtained by fitting the SVM model on the training dataset and optimizing the cost function parallelly (Schölkopf et al., 2007). SVR is an optimization problem by first defining a convex e-insensitive loss function that must be minimized and then finding the flattest tube that contains most of the training instances. Therefore, the risk function (cost function) and constraints are formulated as follows (Khanna, 2015):

$$R\langle w|b|\zeta\rangle = \frac{1}{2}\|w\|^2 + C\sum_{i=1}^{L}\left(\zeta^+ + \zeta^-\right) \tag{6.5}$$

S.t.,

$$\langle w | \varphi(x_i)\rangle + b - y_i \le \varepsilon + \zeta^+_i$$
$$\{y_i - \langle w | \varphi(x_i)\rangle - b \le \varepsilon + \zeta^-_i \tag{6.6}$$
$$\zeta^+_i, \zeta^-_i \ge 0$$

$$i = 1,2,3,\ldots\ldots.L$$

Here, $\varphi: X \to P$ is a conversion of feature space X into a new, more significant dimension space P by using kernel function, as defined below:

$$\langle \varphi(x)|\varphi(x')\rangle = \sum_i \varphi_i(x)\cdot\varphi_i(x') = k(x,x') \tag{6.7}$$

Karush–Kuhn–Tucker (KKT) optimality conditions are prerequisite and enough because the above problem comes to the quadratic family with linear constraints. The dual problem solution of (w and $f(x)$) for support vectors are as follows (John Shawe-Taylor, 2004):

$$w = \sum_{i=1}^{L}\beta_i\cdot\varphi(x_i) \tag{6.8}$$

$$f_{w,b}(x) = \sum_{i=1}^{L}\beta_i\cdot\langle\varphi(x_i)|\varphi(x)\rangle + b = \sum_{i=1}^{L}\beta_i\cdot k(x_i,x) + b \tag{6.9}$$

There are many types of kernels found in several papers to solve regression problems (Steinwart, 2001; John Shawe-Taylor, 2004). Some of the essential kernels are described below:
Gaussian RBF kernel:

$$k(x_i,x_j) = e^{\frac{-\|x_i - x_j\|^2}{2\sigma^2}} \tag{6.10}$$

Polynomial kernel:

$$k(x_i,x_j) = (x_i.x_j + a)^b \tag{6.11}$$

Here, σ, a, b are known as hyper-parameters of the kernel and decide the kernel's behavior accordingly. At last, if SVM is to be used for any regression problem with non-linear separable data, then we have to select a kernel with its best hyper-parameter to regress data accurately.

6.2.3 Grey wolf optimization

The grey wolf (Canis lupus) is a member of the Canidae family. GWO is a newly developed population-based meta-heuristic inspired by the grey wolf pack's social hierarchy and hunting mechanism (Mirjalili, 2014). In a social hierarchy, there are four levels in the wolves' pack (Figure 6.4). There is a descending order of ranking, with alpha being the highest rank wolf, followed by beta, delta, and omega.

The GWO algorithm updates the population solution by iteratively improving the result of the current particle. To finish the hunt, some steps were taken from the wolves' hunting behavior, such as chasing, encircling, and attacking the prey. The equations below can be used to simulate hunting behavior:

Step (1) **Encircling prey**

To hunt, all grey wolves first chase and then encircle the prey (Figure 6.5). This behavior can be simulated in the following way:

$$X(t+1) = X(t) - A.D \tag{6.12}$$

$$D = |C.Xp(t) - X(t)| \tag{6.13}$$

Here,

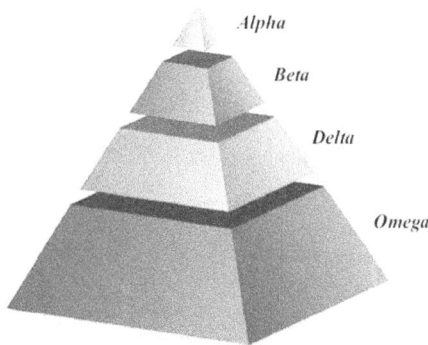

Figure 6.4 Social hierarchy of a grey wolf (Faris et al., 2018).

X (t+1) = updated position of wolf
  X (t) = current location of wolf
  A = controlling parameter
D = distance vector, which represents the distance between the grey wolf
  and the location of prey (X_p)

$$A = 2a * r1 - a \qquad (6.14)$$

Here,
$C = 2*r_2$, controlling parameter
  a = it is the vector whose values decreased linearly from [2, 0]
r_1, r_2 = random vectors whose value lies between interval [0, 1].

$$a = 2 - t\left(\frac{2}{T}\right) \qquad (6.15)$$

Here,
t = current iteration
T = maximum iteration
Figure 6.5 shows how wolves encircle the prey and update their positions accordingly using Eq. 6.12. One solution can update itself to another one using the abovementioned equations. The possible positions of a wolf can be better understood from Figure 6.5. The random components in the formulas above reflect various step sizes and movement velocities of grey wolves.

Step (2)   **Attacking the prey**

After encircling, wolves locate the target and attack to exploit the solution. Since the prey location is unknown at the start, it has been assumed that alpha, beta, and delta wolves know prey location better. Thus rest of

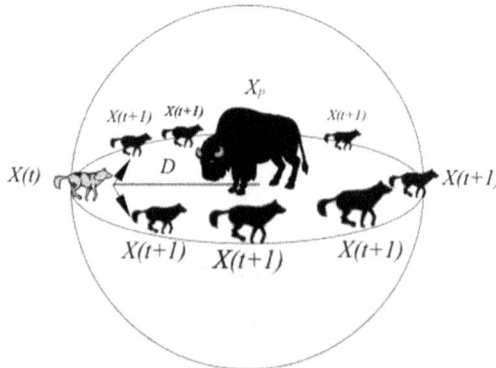

Figure 6.5 Position updating of grey wolves by using distance (D) between prey location (Xp) and wolf from Eq. 6.12 and encircling of prey (Faris et al., 2018).

the wolves update their positions using the location of these three wolves (Figure 6.6). This behavior can be simulated using the following equations:

$$X(t+1) = \frac{1}{3}X1 + \frac{1}{3}X2 + \frac{1}{3}X3 \tag{6.16}$$

Here, X_1, X_2, and X_3 are calculated as shown in Eq. 6.17:

$$\begin{aligned}
X1 &= X\alpha(t) - A1.D\alpha \\
X2 &= X\beta(t) - A2.D\beta \\
X3 &= X\delta(t) - A3.D\delta
\end{aligned} \tag{6.17}$$

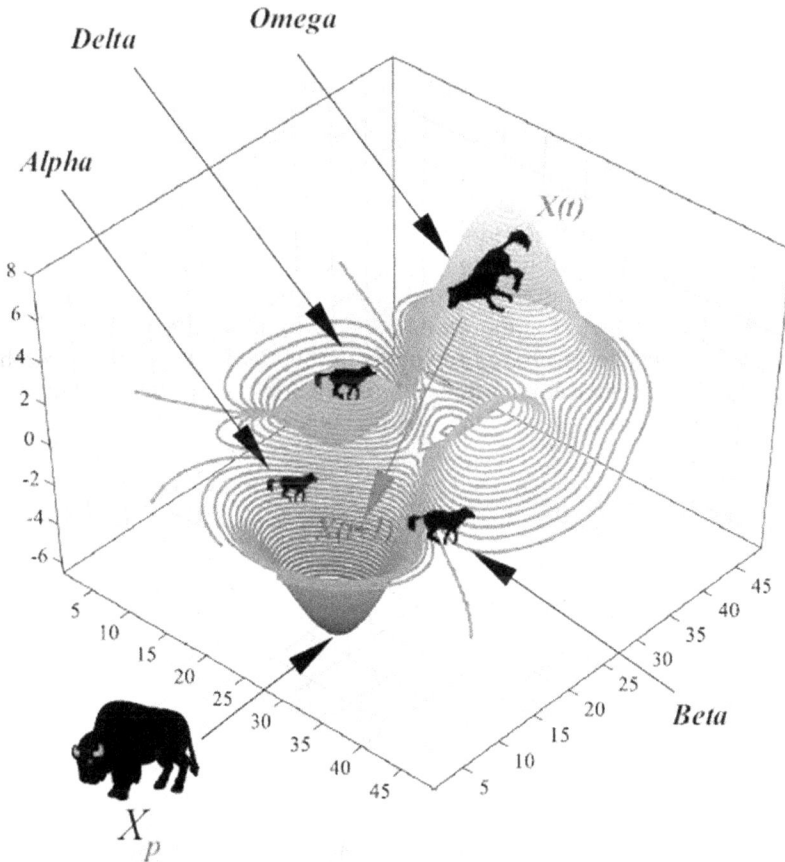

Figure 6.6 Omega positions update according to the location of alpha, beta, delta, and prey.

Here, Dα, Dβ, and Dδ are calculated as shown in Eq. 6.18:

$$D\alpha = |C1.X\alpha - X|$$
$$D\beta = |C2.X\beta - X|$$
$$D\delta = |C3.X\delta - X|$$

(6.18)

During the execution of GWO, it was observed that the first half of the maximum iteration is used for exploration, and the second half is used for exploitation. Exploration aims to explore the search space and find the most plausible zones. In contrast, the exploitation's objective is to exploit the zone and shift the solution toward the optimum global point in the given search space with each iteration. Finally, GWO terminates after exploiting the near solution with a maximum number of iterations as termination criteria.

6.2.4 Modified grey wolf optimization

No meta-heuristic gives the best solution to all optimization problems, according to the no free lunch theorem (Wolpert and Macready, 1997). A meta-heuristic may result better for a set of problems, but on the other hand the same algorithm performs poorly on a different set of problems.

GWO has been modified to align with the search space of complex domains due to the complex nature of real-world optimization problems. Because the GWO has some limitations in its application to real-world problems, some changes have been made in this study to update the mechanism for performance enhancement. In this work, there was a modification of controlling parameter (C) for better exploration and exploitation of search space so that GWO performance may increase. For Alpha (C_1) and beta (C_2), we introduce modified parameter automation schemes that differ significantly from existing approaches (Kundu et al., 2014):

$$C_1 = 0.5 + 0.5^* \left(e^{-\frac{iter}{500}} \right) + 1.4 (sin\ iter\ /\ 30)$$

(6.19)

$$C_2 = 1 + 1.4^* \left(1 - e^{-\frac{iter}{500}} \right) + 1.4\ (sin\ iter\ /\ 30)$$

(6.20)

C_1 and C_2 are control parameters that determine the maximum step size of alpha and beta search agents with the time-varying nature induced by sinusoidal terms:

$$C_3 = \frac{1}{1 + e^{\frac{-0.0001}{T}}} + 2^* \left(\frac{iter}{T}\right)^2 \qquad (6.21)$$

C_3 is the control parameter used for delta wolf in the study. Here, Eq. 6.21 represents the concept of a sigmoid-based acceleration coefficient. It can smoothly convert a linear state into a non-linear state (Tian et al., 2019).

6.2.5 Particle swarm optimization

The PSO is a population-based meta-heuristic in which the population of particles (search agents) and solution updates in each iteration shift toward the optimal solution. In the PSO algorithm, a new population is updated by changing the positions of the earlier particles. PSO is inspired by the social behavior of bird flocking. PSO's search strategy is that "All the birds don't know where food is, but they know how far they are in each iteration" (Olsson, 2011).

The first step in PSO is to create a matrix containing each particle's position vectors Xi (possible solutions to the problem). This matrix represents the position of all particles in the search space. The number of variables to be optimized for a problem defines the particle dimension in search space. The particle's X_i position and velocity V_i are chosen at random. The second step is to calculate the fitness function value for each particle and find the best value among all particles, which is termed the "g_{best} value." All other search particles use the position of a particle corresponding to the g_{best} value to update their location in the search space. The algorithm updates the velocities and positions of all particles using the following two equations (Olsson, 2011):

$$v_i^{K+1} = \omega v_i^k + \varphi_1 \left(g^k - x_i^k\right) + \varphi_2 \left(I_i^k - x_i^k\right) \qquad (6.22)$$

$$x_i^{k+1} = x_i^k + v_i^{k+1} \qquad (6.23)$$

Here,

$$\varphi_1 = c_1^* r_1, \quad \varphi_2 = c_2^* r_2$$

C_1, C_2 = two positive constant
r_1, r_2 = random numbers lie between [0, 1]

The velocity of a particle depends on the following three terms:
The first term contains ω (inertia weight) that affects the previous particle's velocity (v_i^k).

The second term is called a "cognitive term." It is the difference between a particle's best-known position (I_i^k) and its current position (x_i^k).

The third term is called a "social learning term." It is the difference between a particle's global best position (g^k) and a particle's current position (x_i^k).

Initially, a cloud of particles forms over the entire search space to explore the most plausible zones, known as exploration. Further, it begins contracting the specific zone to exploit the best possible solution, known as exploitation. Exploration is thought to have taken place before exploitation (Olsson, 2011).

PSO has been launched in several variants. According to a recent variant, "standard PSO 2011," the value of parameters has been defined as follows:

$$\omega = \frac{1}{2^* \ln 2} \tag{6.24}$$

$$c_1 = c_2 = 0.5 + \ln 2 \tag{6.25}$$

The topology of the swarm (group of particles) describes how particles are linked together to exchange data with the global best. Each particle in the PSO informs only K other particles. Generally, it has three particles that are chosen randomly.

6.3 RESULTS AND DISCUSSION

We conducted a related interrogation in a programming environment to better understand the feasibility and effectiveness of the GWO algorithms. NASA milling dataset was used to check the performance of GWO. It is a public dataset with (167 rows × 13 columns). However, only nine columns were found helpful for our investigation, and the remaining columns dropped. Two columns (Case and Run) were not directly useful in this work, so these columns were removed. Tool flank wear (VB) was a column of an output variable. smcDC (DC spindle motor current) column was found to be highly correlated with the smcAC column. Therefore, it was also removed so that risk of overfitting the model could be reduced. Some rows containing missing tool flank wear values (output variable) have been eliminated. Finally, after data cleaning, we considered only nine input variables to predict tool flank wear value for 145 sample sizes in our experiment. An experiment was performed using the following experimental criteria (Table 6.3).

After data cleaning, the proposed models were executed using initial experimental parameters with 20 population size, 100 maximum iteration, 0.00005, and 100 as lower-bound and upper-bound values of search space, respectively. These parameters are obtained after the number of experiments

Table 6.3 Experimental criteria

Criteria	Value
Population size	20
Maximum iterations	100
Upper bound limit of search space	100
Lower bound limit of search space	0.00005

in which the results were best. A hybrid of meta-heuristic (optimization algorithm) and machine learning algorithm was used to predict tool flank wear value. SVM has been employed as a machine learning algorithm. GWO is used to optimize three hyper-parameters (Regularization factor, gamma, and epsilon) of SVM with minimum cost function value (objective function value). We have listed down the optimized value of SVM's hyper-parameters obtained from MGWO, PSO, and GWO, which are shown in Table 6.4, respectively.

In Table 6.5, we performed a comparative analysis between different hybrid algorithms, as mentioned in the previous paragraph. GWO sometimes may stagnate to local optima, so to improve the performance we have employed a modified version of GWO and benchmarked it with PSO (Robleda et al., 2015) based on the performance metric.

Table 6.4 Optimal hyper-parameter values of support vector machine using MGWO, PSO, and GWO algorithms

S.N.	Solution approach	Regularization factor (C)	Kernel parameter (gamma)	Epsilon (ε)
1.	**MGWO-SVM**	**23.9721**	**0.5546**	**9.5954e-05**
2.	PSO-SVM	57.8773	0.5778	0.014021
3.	GWO-SVM	46.9359	0.6530	0.000121

Table 6.5 Comparative results of different regression models

S.N.	Solution approach	r^2	RMSE	MSE	Execution time (sec)
1.	SVM	48.64%	0.1787	0.0319	2.424
2.	GWO- SVM	94.50%	0.0584	0.0034	43.674
3.	PSO- SVM	95.00%	0.0556	0.0030	41.286
4.	**MGWO- SVM**	**95.30%**	**0.0540**	**0.0029**	**32.349**

From Table 6.5, one can conclude that our proposed hybrid regression model (MGWO)-SVM) predicts tool flank wear value more accurately with 95.30% regression accuracy than other hybrid models. Thus, the first objective was accomplished with 95.30% regression accuracy and less executive time, 32.349 seconds. For better visualization, the following figures depict curve fitting graphs between measured flank wear (observed values) and predicted flank wear values for different hybrid regression models, as discussed previously.

Figure 6.7 shows the fitting of two different curves. The legends in the diagram indicate that the blue curve represents observed flank wear, and the red curve represents predicted flank wear. "Coefficient of determination" (r^2) is a statistical term that describes how well a regression model fits over a set of observed values. The value of r^2 ranges between 0 and 1. Zero indicates the worst fitting of data and 1 perfect fitting of data. From Figure 6.11, r^2 value 0.4864 signifies that the SVM regression model explains 48.64% of the total variation in the dependent variable (Flank wear). Therefore, it can be said that the simple SVM model has proved to be less effective for predicting the data and giving poor fitting compared to other models.

Figure 6.8 shows the result of the GWO-SVM hybrid regression model. From Table 6.5, it has been found that it can fit 94.50% of data. It outperformed a basic SVM regression model.

Figure 6.9 shows the result obtained from the PSO-SVM regression model. This regression model has performed better than previous models in terms

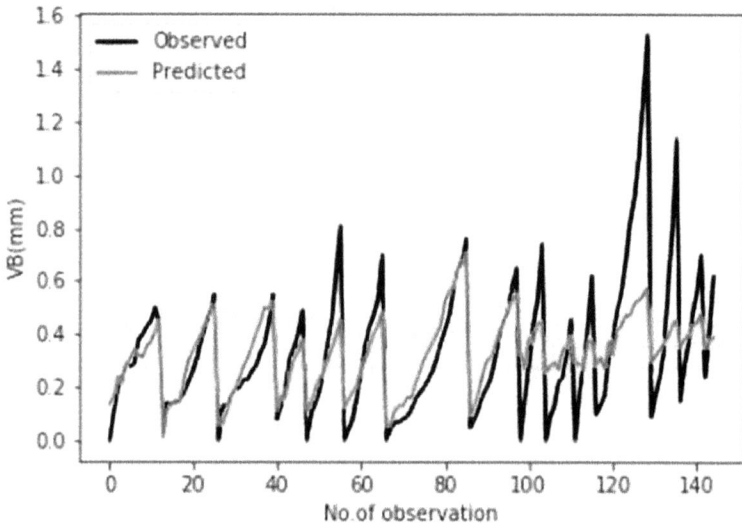

Figure 6.7 Curve fitting graph between observed flank wear and predicted flank wear using SVM.

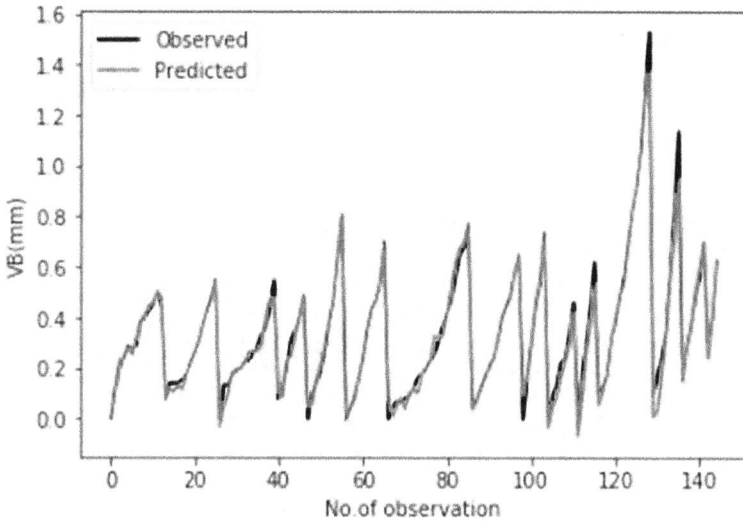

Figure 6.8 Curve fitting graph between observed flank wear and predicted flank wear using GWO-SV.

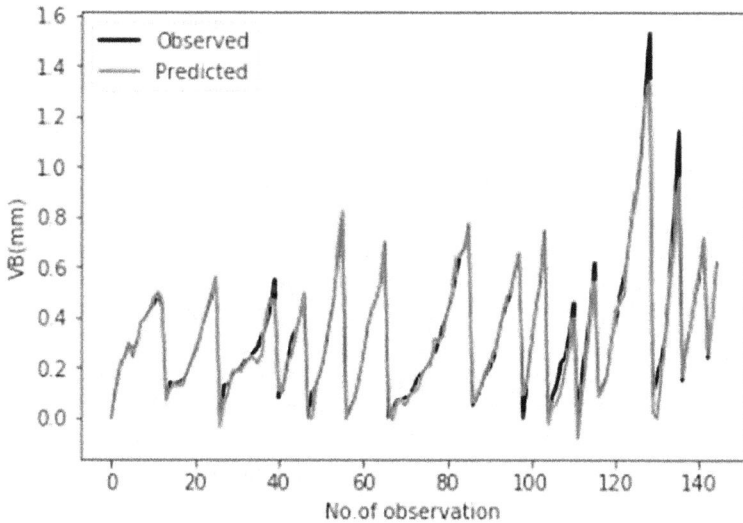

Figure 6.9 Curve fitting graph between observed flank wear and predicted flank wear using PSO-SVM.

of accuracy and computation time. It has demonstrated 95% accuracy in predicting tool flank wear value from actual data, according to Table 6.5.

Figure 6.10 shows the results obtained from the proposed MGWO-SVM regression model. Table 6.5 shows that MGWO-SVM outperformed with

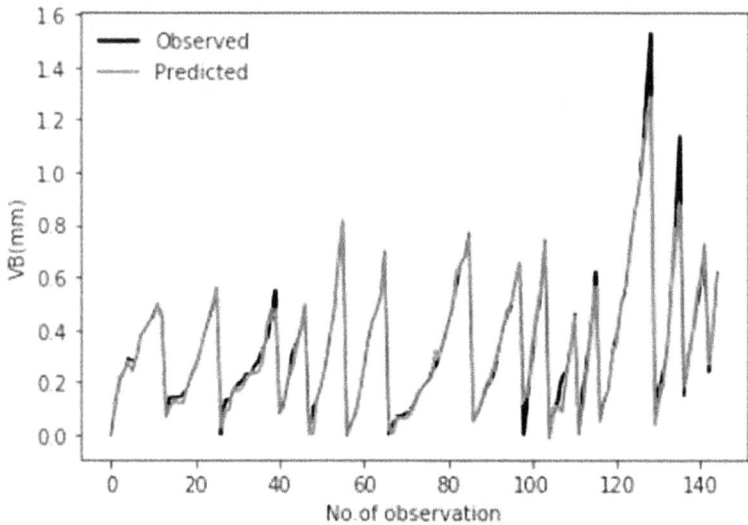

Figure 6.10 Curve fitting graph between observed flank wear and predicted flank wear using MGWO-SVM.

the highest accuracy of 95.30% and the lowest computation time of 32.349. It has proved the ability to fit 95.30% of data accurately. It might be a cost-saving model also due to less execution time. However, due to the less difference between r2 values, it may be hard to differentiate Figures 6.8, 6.9, and 6.10.

Further, our secondary aim was to identify influential input factor which primarily affects tool flank wear to suggest the milling machine's improvement in the future. It has been accomplished by calculating feature weight coefficient values during an experiment in a programming environment. Weight Coefficients are nothing but coordinates of a normal vector (which is perpendicular to the hyperplane). These are relative magnitudes, and the negative sign ('–') indicates the position of the coordinate on the normal vector across the hyper-plane. In Table 6.6, we listed the values of the weight coefficient obtained for different hybrid regression models. The following graphs represent input variables' relative influence on predicted tool flank wear value.

Figure 6.11 depicts the ranking of nine influential variables to predict tool flank wear. The duration of the experiment (time) has been identified as the most significant variable in predicting tool flank wear value. Subsequently, work material and DOC were determined as influential variables. We might suggest better improvements in the milling machine by setting optimal values of these controlling parameters (influential variables).

The experiment started with data cleaning due to many missing values in a dataset; after data cleaning, the proposed models were executed using

Table 6.6 Evaluation of influential variables using feature weight coefficient values

S.N.	Input variable	Weight
1.	Time	0.79
2.	DOC	0.13
3.	Feed	0.04
4.	Material	0.25
5.	smcAC	−0.02
6.	Vib_table	−0.11
7.	Vib_Spindle	−0.08
8.	AE_table	0.03
9.	AE_Spindle	−0.03

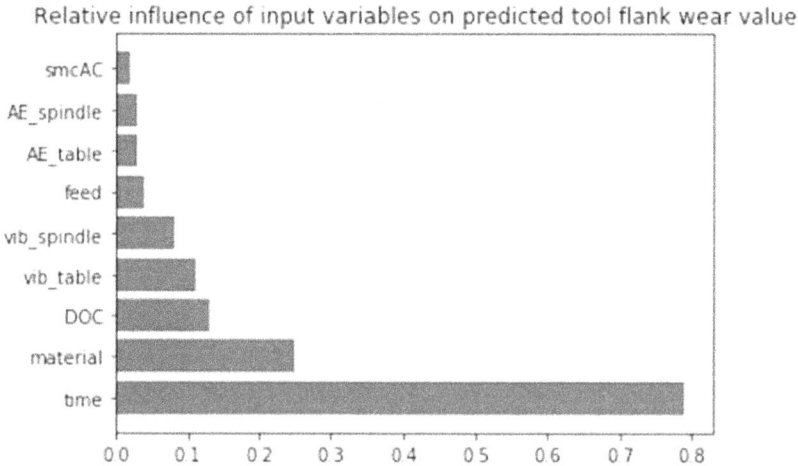

Relative influence of input variables on predicted tool flank wear value

Figure 6.11 Representation of most influential parameters of tool flank wear using MGWO-SVM.

initial experimental parameters with 20 population size, 100 maximum iteration, and 0.00005 and 100, respectively, as lower-bound and upper-bound values of search space. These parameters are obtained after the number of experiments in which the results were best. Optimized values of hyper-parameters are obtained in 3-D search space for the milling experimental dataset, as listed in Table 6.1. These three variables signify the position of the alpha wolf or the position of prey (hunted by an alpha wolf) in search space. Comparative performance for different solution approaches was done in the python language platform and listed in Table 6.5.

The coefficient of determination is a statistical measure that signifies how much the prediction curve fits actual values. Its value ranges between 0 and 1.

Thus, it is '1' for the best curve fitting and 0 for the worst. Finally, tenfold cross-validation is a technique used for calculating the mean value of the coefficient of determination for ten different subsets of the dataset. MGWO)-SVM hybrid regression model has outperformed other algorithms with the coefficient of determination ($r^2 = 0.9530$) and minimum prediction error (RMSE = 0.0540, MSE = 0.0029). These errors are evaluated on a test set with 29 sample sizes. The abovementioned mean errors signify those average values of the difference between actual flank wear value and predicted flank wear value. The significance of R^2 is that it is the proportion of the total deviation independent variable (tool flank wear in our case), best explained by a regression model.

Execution time (or running time) is the time taken to execute all operations of an algorithm. Running time analysis is considered for determining changes in running time if the number of operations increases or decreases in the future. The MGWO was executed in less time than others, within 32.349 seconds. It means MGWO is faster in computation also and a time-saving algorithm comparatively. Hence, it has been proved from the results that our proposed hybrid regression model is superior to others. Additionally, significant decreasing order for a set of most influential parameters is identified as the *duration of the experiment (time), Material, DOC,* and so on. This will aid the milling machine's improvement during the next maintenance by setting values of influential parameters.

6.4 CONCLUSION AND FUTURE WORK

This study presented a new hybrid regression model based on grey wolves' behavior. In the proposed solution approach, a grey wolf pack's social hierarchy structure and hunting mechanism were mathematically modeled. The NASA milling dataset was used to benchmark the proposed solution methodology to other hybrid models in terms of regression matrices (r^2, RMSE, MSE) and execution time of models. Compared to well-known meta-heuristics like PSO and standard GWO, the results showed that MGWO could produce highly competitive results.

From the results, it has been concluded that MGWO performed more accurately in predicting tool flank wear value with less computational time. Due to the simplicity of the model, it allows for predicting tool flank wear value at a low cost for quality assessment in a milling machine and demonstrating how the proposed algorithm can be used to solve other real-world problems.

Furthermore, we have discussed the vital sequence of the most influential factors affecting tool flank wear values (output). The sequence of the three most significant factors was recognized as the *duration of the experiment (time), Material,* and *DOC.* Therefore, the milling machine can be easily improved by optimizing these process variables during the next maintenance.

Finally, this new solution approach can be applied to other machining processes by considering each process's unique characteristics. Future work will contain modifications in the proposed algorithm to enhance performance by introducing new operators, change in population structure and hierarchy of the grey wolf pack, hybridization of two meta-heuristics to take advantage and features of each other, and parallelism to reduce execution time.

REFERENCES

Brugsch, Henrich Karl. *A History of Egypt under the Pharaohs*. Translated by Philip Smith Henry Denby Seymour. London: Cambridge Library Collection - Egyptology, 1881.

Chen, J-L, G-S. Li,S-J. Wu. "Assessing the potential of support vector machine for estimating daily solar radiation using sunshine duration." *Energ Convers Manage*, 2013.

Cristianini, Nello, John Shawe-Taylor. *An introduction to support vector machines and other kernel-based learning methods. Fourth*. Cambridge: Press Syndicate of the University of Cambridge, 2003.

Dibike, Y.B., S. Velickov, D. Solomatine, M.B. Abbott. "Model induction with support vector machines: introduction and application." *Journal of Computing in Civil Engineering* 15, no. 3 (2001): pp. 208–216.

Faris, Hossam, Ibrahim Aljarah, Mohammed Azmi Al-Betar, Seyedali Mirjalili. "Grey wolf optimizer: a review of recent variants and applications." *Neural Computing and Applications (Springer link)* 30, 2018: 413–435.

García Nieto, P.J., J. Martínez Torres, M. Araújo Fernández, C. Ordóñez Galán. "Support vector machines and neural networks used to evaluate paper manufactured using Eucalyptus globulus." *Applied Mathematical Modelling*, 36(12), 2012: 6137–6145.

García Nieto, Paulino José, Esperanza García-Gonzalo, Celestino Ordóñez Galán, Antonio Bernardo Sánchez. "Hybrid ABC optimized MARS-based modeling of the milling tool wear from milling run experimental data." *Materials (MDPI)*, 9(2), 2016: 82.

Goebel, K. "Management of Uncertainty in Sensor Validation, Sensor Fusion, and Diagnosis of Mechanical Systems Using Soft Computing Techniques." P.H.D. Thesis, Department of Mechanical Engineering, University of California at Berkeley, Berkeley, 1996.

Jian, Y.L., T.C. Chun, W.C. Kwok. "Using support vector machines for long term discharge prediction." *Hydrological Sciences Journal*, 51(4), 2006: 599–612.

John Shawe-Taylor, Nello Cristianini. *Kernel Methods for Pattern Analysis*. Cambridge: The Press Syndicate of the University of Cambridge, 2004.

Kecman, V. Support vector machines–an introduction. In *Support vector machines: theory and applications* (pp. 1–47). Berlin, Heidelberg: Springer Berlin Heidelberg, 2005.

Khan, S.M., P. Coulibaly. "Application of support vector machine in lake water level prediction." *Journal of Hydrologic Engineering*, 11(3), 2006: 199–205.

Khanna, Mariette Awad Rahul. *Efficient Learning Machines: Theories, Concepts, and Applications for Engineers and System Designers* (p. 268). Springer Nature, 2015.

Kol, Legát A. *Systémy managementu jakosti a spolehlivosti v údržbě.* ČSJ, Praha, 2007.

Kundu, Rupam, Swagatam Das, Rohan Mukherjee, Shantanab Debchoudhury. "An improved particle swarm optimizer with difference mean based perturbation." *Neurocomputing (Elsevier)*, 129 (2014): 315–333.

Mesut, C. "Estimation of daily suspended sediments using support vector machines." *Hydrological Sciences Journal*, 53(3), 2008: 656–666.

Mirjalili, Seyedali. "Grey Wolf Optimizer." *Advances in Engineering Software*, 69, 2014: 46–61.

Moghaddamnia, A. Ghafari, M. Piri, J.D. Han. *Evaporation estimation using support vector machines technique.* World Academy of Science, Engineering and Technology, 2008.

Nieto, P.J.García. "Hybrid modelling based on support vector regression with genetic algorithms in forecasting the cyanotoxins presence in the Trasona reservoir (Northern Spain)." *Environmental Research*, 122 (2013): 1–10.

Olsson, Adrea E. *Particle swarm optimization: Theory, techniques and applications.* United States: Nova Science Publishers, Inc., 2011.

Orosz, Tamás, Péter Sőrés, Dávid Raisz, Ádám Z. Tamus. "Analysis of the Green Power Transition on Optimal Power Transformer Designs." *Periodica PolytechnicaElectrical Engineering and Computer Science*, 59(3) (2015): 125–131.

Panigrahi, Bijaya Ketan, Yuhui Shi, Meng-Hiot Lim. *Handbook of Swarm Intelligence: Concepts, Principles and Applications.* New York: Springer Science & Business Media, 2011.

Robleda, P. J. García-Nieto, García-Gonzalo, J.A., Vilán Vilán, Segade. "A new predictive model based on the PSO-optimized support vector machine approach for predicting the milling tool wear from milling runs experimental data." *International Journal of Advanced Manufacturing Technology*, 86 2015: 1–12.

Scay, John. *Introduction to Manufacturing Processes .* New York: McGraw-Hill, 1999.

Schölkopf, Bernhard, Alexander J. Smola, Francis Bach. *Learning with Kernels: Support Vector Machines, Regularization, Optimization and beyond.* Cambridge: MIT press Cambridge, 2007.

Steinwart, Ingo. "On the Influence of the Kernel on the Consistency of Support Vector Machines." *Journal of Machine Learning Research* 2 (2001): 67–93.

Tian, Dongping, Xiaofei Zhao, Zhongzhi Shi. "Chaotic particle swarm optimization with sigmoid-based acceleration coefficients for numerical function optimization." *Swarm and Evolutionary Computation (Elsevier)* 51 (2019): 1–16.

Wolpert, David H., William G. Macready. "No free lunch theorems for optimization." *IEEE Transactions On Evolutionary Computation (IEEE)* 1 (April 1997): 1–16.

Chapter 7

The energy consumption optimization using machine learning technique in electrical arc furnaces (EAF)

Rishabh Dwivedi, Ashutosh Mishra, Devesh Kumar, and Amitkumar Patil

Malaviya National Institute of Technology, Jaipur, India

CONTENTS

7.1 INTRODUCTION

Steel is an alloy of iron manufactured using scraps or iron ore which is found in a wide range of applications. To make multiple grades of steel, impurities like nitrogen (N), silicon (Si), phosphorus(P), sulfur(S), and excess carbon(C) (impurity and alloying element) are removed from iron and elements like manganese (Mn), nickel (Ni), chromium (Cr), and vanadium (V) are added for alloying (Urbański, 1974).

Steelmaking has been around for ages, but wasn't widely commercialized till the 14th century. Some manufacturing techniques have emerged and flourished, while others have become obsolete and been abandoned. All of these shifts were linked and influenced one another. If this method is studied apart from the evolution of making steel, the comprehension of electric steel manufacturing development and possibilities would be incomplete (Toulouevski & Zinurov, 2013). The crucible technique was an old steelmaking method. Steelmaking became a significant business in the 1850s and the 1860s because of the Bessemer and Siemens-Martin processes. The earlier-mentioned processes have gone extinct. New techniques like blast furnaces use pig iron, and electrical arc furnace (EAF) use scrap or direct reduced iron as a raw feed material. Technological development led to oxygen's chemical energy use in EAF.

The EAF makes molten material from scrap materials and alloys, purified to a predetermined steel grade using downstream operations (Saboohi et al., 2019). A bucket with raw materials is used to feed the furnace just at the start of the procedure. When the ceiling of the furnace is covered, and after the electrodes are turned on, the melting period starts. This step continues until the materials have liquefied enough to allow for adding a second bucket of raw materials (Argiolas & Bacchetti, 2020). After the subsequent charging step, there is yet another melting stage. During both melting phases, propane and oxygen burns assist in melting cold regions which would otherwise prevent even melting. A refining step begins after most of the charged feedstock has been dissolved, wherein the steel is adjusted to a preset target composition. In conjunction with lanced oxygen, elements like silicon and carbon are added. The exothermic chemical reactions caused by oxide heating of the steel lower the quantity of electrical energy (EE) required. The steel is then tapped into the ladle and transferred for other processes in the final step. Any necessary furnace arrangements are completed before another heat (Kovačič et al., 2019).

Today's EAFs would have looked unthinkable 20 years ago. The melting time in the most advanced furnaces (capacity of 90–140 t) has been decreased to 30–40 minutes thanks to an astonishing number of advances. Electric power usage was cut in half, from 640 to 350 kWh/t, but hourly kept rising sixfold, from 50 to 250 t/h. The EE accounts for 50% of the total energy usage per melt (Saboohi et al., 2019). The consumption of electrodes was cut in half. In the coming years, such performances will likely be the standard for most steelworks.

The two essential operations in modern furnaces are melting solid scraps and liquid bath heat. As a result, these elevated processes account for today's burner productivity. Heat should be generated via electrical or chemical energy and afterward transferred to the areas of the material charge or liquid bath to start these activities (Carlsson et al., 2019). Heating technologies, furnace design, and other EAF machinery change at breakneck speed (Kovačič et al., 2019). The latest technological solutions are presented and heavily advertised annually. Manufacturers are having difficulty navigating the deluge of new technologies.

A conventional EAF is generally activated in accordance with predetermined melting profiles that rely primarily on input energy and worker experience (Saparrat et al., 2020). The characteristics maximize EAF performance; however, the established profiles do not account for changing EAF conditions. Effective EAF actuation is critical for achieving optimal slag properties, which decrease energy usage and disturbance, shield the sidewalls and water-cooled panel, and assist in the ideal end-point metal composition. Due to a shortage of understanding of the EAF process, event initiation times, such as filling, carbon input, and oxygen lancing, frequently depart from optimum periods, resulting in reduced EAF efficiency (Carlsson et al., 2020a; Murua et al., 2020).

As every manufacturing industry is transitioning toward sustainability by way of prevention from hazardous environmental effect of processes, the utilization of EAF to produce stainless steel has increased; this led to the studies in the field for using reduced energy consumption and raw material for the process.

Additionally, it comprehends supply dependability, service quality, voltage, current, source quality, and the effectiveness of using electrical power. Maintaining the sinusoidal waveform of the electric supply voltage magnitude at rated voltage frequency and magnitude is referred to as power quality. Steel manufacturing using an EAF requires a lot of energy; thus, improving efficiency and power quality is a significant concern for every stainless-steel industry. One of the essential factors is electrical power consumption, along with slag foaming, which is a crucial component to ensure quality and performance in an EAF steelmaking process (Carlsson et al., 2020a). The higher version of the EAF minimizes electrode consumption, tap-to-tap time, and proper utilization of chemical energy (Tomažič et al., 2022).

There can be several features that affect the power consumption in the furnace. We will discuss some of them and the improvement, optimization, or predicting approach using ML techniques. Scrap melting for steel production requires nearly 400 kWh/t depending on grade and input scrap. Waste is generally based on operator input being charged into the furnace based on several things, such as availability, cost, grade, and the practices that have been carried out for longer (Murua et al., 2020; Tomažič et al., 2022). A machine learning–based mathematical modeling can be designed for scrap charging to optimize EE. The production cost varies linearly with graphite electrode consumption, and energy optimization is done using the linear programming problem. That linear programming problem can be solved using linear quadratic regulation, which provides feedback from closed-loop solutions and is more robust than open-loop solutions.

There is a different statistical model based on various EAF parameters that have been used to detect the power consumption of the furnaces. Linear models such as multivariate linear regression or mean EAF values started in the early 1980s; they used prevalent values and tried to set up a relation among them. The matters considered are the weight of scrap charged, fluxes, alloys, etc. Still, they had not regarded critical parameters such as slag, carbon injected, etc. and so the results were quite vague, such as a range of 380 to 600 kWh/t. Multivariate linear regression and coefficient did not yield any satisfactory result, so a different linear model was used, called partial least square regression, which has a unique approach. It uses two machine learning (ML) models to predict the per ton usage of EE (Carlsson et al., 2020a; Murua et al., 2020). It sets the relationship between the yield of liquid metal and scrap metal. The authors considered it a non-reliable method. As the studies evolved, other techniques were used instead of linear, such as artificial neural network (ANN), random forest, deep neural network, support vector machines, and decision tree (Carlsson et al., 2020a).

ANNs and neural network theory have been implemented on EAF for different purposes. These techniques were earlier used in measuring and predicting temperature to protect the furnaces' lining. With the help of the decision tree and random forest and chaos theory, ANN has been used to indicate the actual temperature because the temperature is very high inside the furnace due to radiation, and there have been complaints of defection on the electrical temperature measuring devices (Murua et al., 2020; Tomažič et al., 2022).

Machine learning and statistical models are used to predict many behaviors of EAF, such as the arc length, voltage, current, and other parameters. There are many fields still to be observed. Many uncertainties in the furnace need to be explored because a number of things are still being done. After all, it's been done for a long time, and experimenting with all permutations is impossible (Argiolas & Bacchetti, 2020; Murua et al., 2020). The cause why ML is helpful in this field is that a considerable number of possible combinations exists for different grades, and to carry out experimentation continuously would require a large amount of money, so it is good that we can now predict the behavior using previous data and can test it for the betterment of the performance.

In Section 7.2, "Literature review," we will take an overview of the plant using literature and will also see the evolution that took place for the improvement in EAF. Section 7.3, "Methodology," explains data collection and analysis procedures. Section 7.4, "Result and discussion," will be followed by Section 7.5, "Conclusion limitations and future scope."

7.2 LITERATURE REVIEW

In this section, in general, we discuss literature survey throughout the timeline when it was defined as a detailed description of techniques used for improvement, machine learning techniques, and their progression in the field.

First, we searched the Scopus database for "steel industry" and "machine learning," through which we received numerous documents. It was divided into following four broad categories: raw material, related to melting, hot rolling, and cold rolling.

The steelmaking industry uses scrap steel to reproduce steel of the required grade (Shyamal & Swartz, 2019). It was done because it is cost-effective and energy-saving rather than using iron ore. Discarded metals reach steel mills in large, well-equipped trucks, making it challenging to visualize goods. Often heavy residues are placed on the ground, and small and mixed residues are placed on top for load improvement. That makes it difficult to verify the type of scrap shipped and the type of material under load. It is intended that the classification process will be improved with time, as well as reliability to reap financial benefits and improvement of the quality of the separation and disassembly of metallic scrap that will help separate the metals. The application in the steel industry was considered both in segregation removal of metal fragments and in preparation of scrap production, to be used in producing stainless steel (Todshki & Ranjbaraki, 2016).

ML techniques can be used for programming mathematical and statistical models for extensive data processing to classify and predict scrap behaviors and chemical composition (Huang & Han, 2008). Hyperspectral cameras can be another method of sorting raw material. We can also use a robot-based separator to reduce the laborious task (Liu et al., 2019). There are around 3,500 grades of a steel ML model that can be used to separate raw material based on composition (Carlsson et al., 2020b). They can be used further to classify raw material or scrap optimization, which reduces the time taken to produce a specific grade using a certain kind of raw material. The use of ML classifiers in raw materials can provide us with the combination of raw materials, which will lead to specific chemistry so that we can reduce the input of costly pure metals that are the input in steel. Steel manufacturing unit using an EAF requires a lot of energy; thus, improving efficiency is a top concern for every steel industry. One of the essential factors is electrical power consumption, along with slag foaming, which is a crucial component to ensure quality and performance in an EAF steelmaking process. Within the higher version of the EAF, it can be achieved by minimizing electrode consumption, reducing tap-to-tap time, and properly utilizing chemical energy.

Several parameters can affect the arc furnace's power consumption and quality. We will discuss some of them and the improvement, optimization, or predicting approach using ML techniques. Scrap melting for steel production requires nearly 400 kWh/t, depending on grade and input scrap (Bai, 2014). Scrap is generally based on operator input being charged into the furnace based on several things, such as availability, cost, grade, and the practices that have been carried out for extended periods. An ML-based mathematical modeling can be designed for scrap charging to optimize EE (Shyamal & Swartz, 2018). The production cost varies linearly with the consumption of energy and graphite electrodes. The optimization is done using the linear programming problem. The linear programming problem can be solved using linear quadratic regulation, which provides more robust feedback than open-loop solutions.

Hot rolling is a process in which the thickness of the slab (solid cuboid structure) is reduced to increase its length. In hot rolling, the slab is heated above recrystallization temperature and then is passed simultaneously through several rolls. Like any process, several defects occur, and to counter this, ML can be used; the research is still proceeding in this direction. A large number of data is generated on such machines using programmable logic control (PLC) and sensor data through which we can use ML to predict machine failures to avoid breakdown maintenance and increase the productivity of the process.

Rolling force is the primary factor affecting the thickness of the strip. Given the complexity and discontinuities of the rolling Machine, as well as several influencing elements, the classic rolling force forecast model's analysis must frequently be streamlined and theorized. As a result, the numerical method is inadequate, and there is a discrepancy between the estimated outcomes and the actual operating conditions. Rolling force prediction in this chapter, using the techniques "genetic algorithm (GA), particle swarm optimization algorithm (PSO), and multiple hidden layers extreme learning machine (MELM) is proposed, namely." Rolling force is monitored constantly because the strip's thickness depends on it (Liu et al., 2019).

One of the most severe faults during a hot rolling process includes camber, which is described as the extent of curvature of a metal plate. The camber, indicated by the bending of the metal sheet that occurs in the rolling process of a steel production, causes productivity and quality problems. The mathematical models traditionally used in the prediction of camber do not take as many variables as the ANN takes so that they yield a more accurate result than the mathematical model (Kim et al., 2018).

EAF optimization approaches for the electrical power reduction can be classified into four groups:

  I. Linear programming
  II. Predictive model control
  III. Linear quadratic regulator (operational optimization)
  IV. Other approaches (genetic algorithm (GA), commercial software)

Mixed integer linear programming (MILP) is a framework for industrial facilities that describes the process as a network of nodes (anti-anti) connected by materials and energy flows. The advantage of this technology is that it allows for a simultaneous depiction of the entire industrial system and optimization of the system as a whole rather than just particular subprocesses (Richards & How, 2005). The usage of torn scrap is maximized in the best solution due to the reduced exergy efficiency of this scrap grade (−50 kWh/t compared to "regular" scrap). Usually, the max amount is crushed scrap is constrained by steel grade's quality limits (chemical analysis). Because of their greater exergy efficiency (+80 kWh/t compared to "regular" scrap), the use of hot briquetted iron/direct reduced iron (HBI/DRI) is minimized (zero consumption).

A comprehensive non-linear space vector model has been created using an EAF model from the literature. The model is composed of 14 non-linear ordinary differential equations (Bekker et al., 1998). The 11 factors that make up the makeup state are divided into three categories: one category consists of steel scrap and solid melt slag powder. The second category includes liquified stainless steel and contaminants in the steel, such as carbon (C) and silicon (Si), as well as molten liquid slag and pollutants in the slag, such as SiO_2 and FeO. The third category consists of gaseous phase elements. Carbon dioxide (CO_2), carbon monoxide (CO), and nitrogen (N) are the three gases. The final state factor may influence the temperatures of the solid and liquid groups and the relative furnace pressure.

Short-term factors like furnace pressures and off-gas content, as well as long-term variables like tap-to-tap intervals and pollution levels, can be used to quantify management objectives. These objectives, which are stated in terms of the long variables, must be converted into short-term variables that can be measured on a plant. In just one plant, there are quantifiable disturbances such as arc power, oxygen flow, slag, and the DRI addition rate (Bekker et al., 1999). DRI is introduced through a duct to reduce the temperature when the furnace's temperature rises too high. Temperature is controlled by adding DRI to the stove. EAF cannot be completely sealed; pollution is always a possibility. Gas that has been contaminated with dust can leave the furnace through one of the vents if the furnace's negative relative pressure isn't high enough or through any unsealed openings. The gas includes a high concentration of carbon monoxide at a high temperature, posing a severe safety risk. As a result, the EAF is frequently at negatively operated relative pressure, resulting in energy loss because the off gas extracts a substantial amount of heat. Improvised control would have allowed the negative relative pressure to be consistently low, resulting in less energy use (Bekker et al., 2000).

The majority of industrial control programs are designed to boost profits. They are accomplished by lowering the cost of manufacturing of a specific product, providing value to the products at no added expense, or mixing the two. On the other hand, most control design goals are expressed in improved terms of regulations, following setpoint, and disturbance rejection rather than economic concerns. It is especially essential in multivariable processes because interactions might invalidate profitability assumptions based on relationships with single-input–single-output (SISO). A method is described for incorporating the economic elements that affect the costing of a multivariable process directly into the controller design, guaranteeing that the steel plant is regulated in the most cost-effective manner possible (Oosthuizen et al., 2004).

Model predictive control (MPC) is a standard APC algorithm. The majority of the time controller design focuses on improving regulation, operating point obeying, and on external disturbance rather than process economics. Nevertheless, economic concerns were included as the core of the control system in an MPC's objective functions. MPC is suitable for controlling the numerous process parameters of commercial significance, according to numerical simulations of an EAF (Bekker et al., 1999, 2000). The financial performance of such an EAF with MPC administration was assessed, and the running cost was significantly reduced. The steel temperature and % carbon, generally adjusted manually in the burner under research, are the critical controllable factors in this research. Only one controllable factor at a time can be considered economic inquiry, which is a flaw. This univariate strategy results in overstated profitability since the financial advantage is inflated when manipulated variable and their efficiency functions are dependent (Zhou & Forbes, 2003). The objective of this study is not to build a case for using MPC on an EAF. It was performed using the referenced source. This research aims to build on previous works by demonstrating that when undertaking an economic evaluation of a controller, it will be more appropriate to account for the interdependence of controlled variables.

The optimal control design model's goal differs from the current research. For example, in (Bălan et al., 2007) predictive control based on an adaptive model was devised to follow the predefined trajectories. Preset trajectories design was not mentioned. In (Boulet et al., 2003), a proportional integral derivative (PID) controller was devised to keep the electrodes' energy usage constant. A model was created by (Çamdali & Tunç 2002), but no debate on how to regulate it took place. In (Huang & Han, 2008), a model was presented to get the highest power input to the melting process. Minimizing electricity consumption was not a priority in any of these projects.

The model adopted by Bai (2014) is a linear time-invariant system that appears to perform well. Appropriate control values are computed in the second stage to lower production costs. The solution is a linear programming problem since the actual manufacturing cost is linear in energy and carbon electrode usage. One of the drawbacks of linear programming is that it uses an open-loop system input architecture. Although it performs best in the lack of model errors and noise measurement, its effectiveness in practice cannot be guaranteed due to model ambiguity and parameter variations. This chapter's significance is to demonstrate that, thanks to the EAF's distinctive shape, the linear programming issue may be handled using the linear quadratic regulation (LQR). As a result, it is more reliable than an open circuit solution. The suggested fix for an arc furnace's best control strategy creates a closed-loop robust system that lowers the cost of EAF steel manufacturing. This research focuses on seven essential factors and linear models. Other minor variables, as well as possible non-linear, should perhaps be examined in the hereafter. The cost of production is predicted to be further decreased by combining all components.

The EAF is a chaotic furnace due to which the linear model fails to predict the variability, so the evolutionary process solves the problem of both constrained and unconstrained optimization using a GA. The procedures of picking the equation's format and finding the factors were the most commonly employed. Because of the enormous number of concurrent physical and chemical phenomena needing a complicated mathematical representation, using simulation based on a system's physical chemistry is challenging. As a result, an attempt was made to adopt an approach based on random calculus. The GA methodology was employed to find the accessible statistical equations characterizing electric power usage in the arc furnace (Czapla et al., 2008).

ANNs are a viable solution to several faculties connected to EAF functioning, including temperature prediction (Kordos et al., 2011; Mesa Fernández et al., 2008) and electrical load modeling (Chang et al., 2010; Wang et al., 2005).

(Olabi et al., 2006) used a training algorithm ANN using the Taguchi technique to design and determine the best heating rate, which is laser powered, and the focal location for CO_2 pinhole laser-based welding of intermediate carbon steel butt welds. Until the ANN absorbs the data for training, the back-propagation algorithm's goal is to diminish the discrepancy between actual and expected outputs. The signals are sent forward using the back-propagation process, and the errors are transmitted backward. Unlike earlier research, the goal of (Gajic et al., 2016) work was to use feed-forward neural ANNs, namely the multilayer perceptron (MLP), to predict the relationship between particular EE usage and liquid steel chemical properties.

To train a neural network, select an appropriate encoder and set all weights so that the difference between the desired and actual outputs is as little as possible. The error component of both the consequences must also be computed during the training phase, which shows how the error varies as each load is increased or lowered significantly. The back-propagation algorithm technique is the most generally used approach for calculating optimized weights traveling from layer to layer in the opposite directions of how activities pass across the network. It uses first-order methods, including the gradient descent method, to optimize the values in a recurrent operation. The ideal solution is found substantially faster using quasi-Newton techniques that leverage second-order derivatives. The load was re-equipped in a combination of the inverse Hessian matrix and the negative gradient vector, and the second-order results were generated in a Hessian matrix, H. The calculation of such a Hessian matrix within all second-order derivatives, on the other hand, takes a long time. To boost speed, an approximation to the Hessian matrix is applied. The Broyden-Fletcher-Goldfarb-Shanno (BFGS) algorithm (Shanno, 1970) is utilized in this study as being the most efficient method of computing load updates. It can be seen from the review of literature that very few articles have been published on application of ML in the steel industry. Even smaller number of papers are available regarding the melting process because we cannot see what's going on inside the furnace due to its high temperature, which is impossible to control. Even then, there have been several developments related to them.

7.3 METHODOLOGY

Steelmaking is an ongoing process that includes phases like material charge, melting, casting, and liquid tapping. Preparatory work, charging, melting, purification, prolonged refining, and tapping are a few smaller tasks that make up the melting process (Carlsson et al., 2019). Figure 7.1 depicts the methodology used for this study. Firstly, understanding of key elements and working of the EAF was very essential; hence, the Scopus database was searched for available literature related to EAF and steel industries. After a thorough understanding of the processes and the key elements and process measures of EAF used in the steelmaking process, a brainstorming was done with the industry experts of an Indian-based steelmaking plant to understand the various forms of power and heat used for the melting process during EAF operation. The experts' input was very critical in terms of

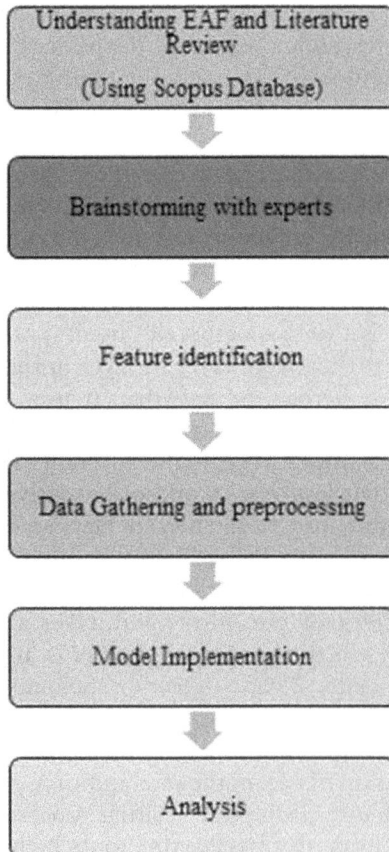

Figure 7.1 Research methodology (data analysis step-by-step).

feature identifications of various heat ratings used for an EAF process, and the feature identification was done with the help and input by the industry experts The feature identification is critical for a data-driven approach such as ML, and it was decided that for a particular grade of steel the collection and preprocessing of data related to the various heat ratings used for a particular grade of steel was done after the feature identification. Thereafter, ML models were implemented on the dataset and implications were drawn following a thorough analysis of the data.

7.3.1 Data-driven model for power consumption in EAF

We created a complete model that uses ML approaches to develop practical solutions for real-world energy challenges to examine energy use in an EAF (Figure 7.1). Understanding the components of any system and its conceptual and functional links is the first phase in evaluating its effectiveness. An energy system for all production systems is "all components relevant to the extraction, transformation, distribution, or use of energy" from a functional standpoint. The essential characteristics of the energy consumption should be determined using feature recognition extracted from the segmented relationships among the constituents of the power system when considering the essence of the energy trouble, assembling the dataset (data collection and preprocessing), even if the modeling step's feed is complex. Fortunately, the increased use of data providers has resulted in a vast volume of data in the corporate data center (Mosavi et al., 2019), speeding up the data collection process. There are three stages in the modeling process. The first stage tries to create EAF energy predictors depending on the industry and manufacturing processes. Classification algorithms based on data extraction use statistics and ML to provide planning abilities. The analytics model skeleton should be selected depending on the essence of a function approximation to attain the highest generalization ability. There are a variety of models available, ranging from traditional models to ANNs and deep learning models, which ensure EAF energy's ability to anticipate various energy consumption. Another function of statistical models for EAF is to indicate the importance of characteristics and their commitment to set the goal value.

Using this helpful result, departmental managers can improve the accuracy of their judgments and align with operational and manufacturing goals. In the second step, descriptive ML models describe what is truly going on within the operations of the under-researched scenario. K-Means Clustering methods help data scientists to find prevailing grouping in observational data evaluated across variables as early analytic tools. The use of descriptive approaches has two significant advantages. On the one hand, differentiation of the decision area leads to a better knowledge of the situation.

On the other hand, they might generate production scenarios based on previous operational data, which would be a valuable tool in planning activities. It implies that situations are created by picking points from a data collection in the same scheme. The choice, along with its associated chances, is a decent estimate of the actual set's distribution function (Kaut, 2021). The third stage, the prime focus of power optimization, uses the results of an illustrative model from second location. Then we try to decide on the operating procedure based on the result. The model's outcome helps operations managers to develop procedures that lead to the most effective manufacturing scenarios.

7.3.2 Understanding key elements affecting EAF's electricity consumption

Steelmaking is an ongoing process that includes phases like material charge, melting, casting, and liquid tapping. Preparatory work, charging, melting, purification, prolonged refining, and tapping are a few smaller tasks that make up the melting process (Carlsson et al., 2019). Furnace, the primary user of electric power in the steel industry, is interconnected with different aspects of the manufacturing process. As illustrated, the scrap kinds delivered to the furnace significantly impact power usage since they directly reflect the input substance's specific gravity and weight and the processing time. Furthermore, the occurrence of exothermal processes would affect the power consumption depending on the metallurgical features of varying scrap kinds (such as the presence of Al, C, Si, and other elements in the input furnace material) (Carlsson et al., 2019). A melting process's components, unavoidable in this procedure, should also be evaluated. Exothermal reactions, in particular, will lead to lower electricity consumption due to oxygen lancing. According to the working group, the condition of the technology used for the process is another potential variable determining electricity consumption and increasing the age of the ladle and crucible since their previous lining could affect the melting process and, as a result, the amount of electricity used.

7.3.3 Influencing feature identification for modeling

After evaluating the issue with the influential aspects, a features pool was created for each component. Perhaps the most significant features were chosen by an advisory committee based on how frequently they were used. Scraps categorization is standard in stainless-steel production facilities that use recycled scrap to measure productivity. Because similar scrap kinds have similar melting efficiency materials with similar features, this activity helps the operative impact on the production process. Evaluated for various and the concentration of scrap types are frequently used to categorize scrap (Carlsson et al., 2019). However, there is no hard-and-fast rule, and steel producers typically rely on inside regulations based on their particular business situation. When we apply the K-Means Clustering, we use power and oxygen as the classifiers based on which clustering is done. We also analyze the principal component to see which factor impacts the modeling behavior.

7.3.4 Creating the heats' dataset

The dataset-building effort is strenuous but worthwhile due to the procedure's intricacy and the steel sector's rigorous working circumstances (Shyamal & Swartz, 2019). We discussed with the company's operations people to record extra data relating to the stainless-steel manufacturing method in addition to the regular, typically recorded database to build the input parameters for the model. The data collection began depending on the attributes chosen and lasted six months of a specific grade during early working shifts. The original dataset was built using two sources of data: the plant's official databases on the server platform and the shift supervisor's report. The apparatus information in the database was used to create conditional variables. Melting compounds were culled out of each heat's shift data. The effective data was individually recorded using the magnet crane's weighting mechanism for each load out from the scrap zone. The furnace's surveillance panel also calculated the electricity used in each heat. The dataset was prepared for the modeling process using two different data preprocessing and treatment procedures (domain-specific approaches and statistical enumeration methods). Due to incomplete, inappropriate, or out-of-range data, 39 records were eliminated from 670 explicit heats' records.

The scatterplot depicts that the operating range for a particular grade JT is not clearly defined, which can be due to a number of reasons. It also depicts that a large chunk of data can be seen between the power range of 17,000 to 19,000, but some heats are significant between 16,000 and 17,000. From this, we can make out that if we make a standard operating procedure, we can narrow the operating range (Figure 7.2).

A screen plot depicted the relative merits of the major components. The weight of the different segments was represented in the screen plot; from this, we can make out that there are three weights with a non-zero value. The most significant are the components 0 and 1, which considerably affect the model (Figure 7.3).

The loading for the top components is shown in the plot. The loadings of components of the principal component analysis (PCA) show the same sign, which shares the factors listed, such as power, total oxygen, liquid metal (LM), tap-to-tap time, etc. The second component talks about the power, and the fourth one talks about the total oxygen composition.

Clustering techniques divided the data set into four clusters, as we can see. This cluster depicts that we can obtain a result of our requirement using a particular set of variables. For instance, we can see that cluster 0 uses energy between 17,000 and 18,000, cluster 1 uses energy between 18,000 and 19,000, cluster 2 uses power more significant than 19,000, and cluster 3 uses energy less than 16,000. We will focus on cluster 3 because it uses very

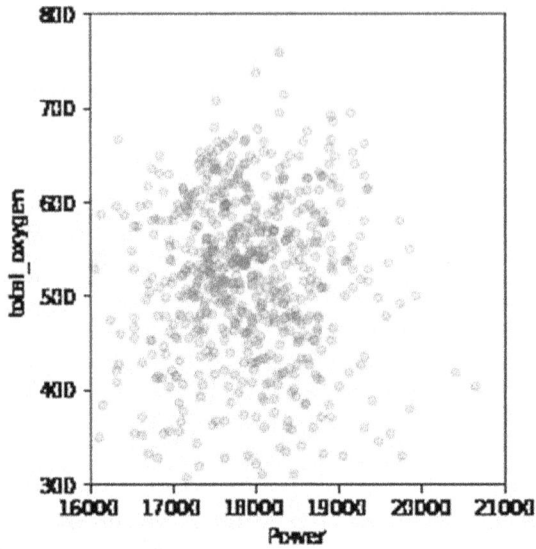

Figure 7.2 Scatterplot for power and total oxygen consumed in heat.

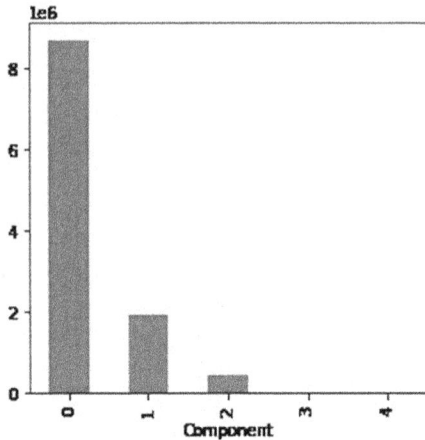

Figure 7.3 Most prominent components.

low energy, and if we design our operating procedure used in cluster 3, we can significantly reduce the energy consumption. In single heat, we can save energy in 1,000s of units. This process can be carried out for different grades, yanking similar results. If we can save energy at this rate, it would significantly affect production and even considerably impact the environment. Using this, we can also study the cause of greater energy consumption. Eventually, we can compare clusters for better result optimization.

There were 631 heats in the dataset, out of which cluster 0 contains 140, cluster 1 has 292, cluster 2 has 71, and cluster 3 has 128. Analyzing each cluster, we notice that specific raw material is present in all the heats, be it any cluster. These raw materials are used to ensure that the heat contains a specific metal required in the process. For example, ferrochrome is required for chrome recovery; similarly, ferronickel is needed for nickle recovery. These raw materials are cheaper than pure metal, so it is more feasible to use these materials. The more profound analysis of the cluster suggests that for grade 304 if the stainless steel 300 is used for the raw material, then power consumption is less. If we analyze the clusters, we know that as the average composition of SS 300 decreases in the charge weight, the consumption also increases. It happens because 300 series material has a similar chemical composition, so excess charging is not required.

7.4 RESULT AND DISCUSSION

The K-Means Clustering for each cluster was used to identify the power usage of EAF, while using cluster centroids as the figure's input parameters and operational conditions. The proportional benefits of several clusters compared to one another are shown in Figure 7.4. The comparative use of scrap kinds was evaluated based on respective proportions in the total chargeable scraps to explore the processes in different cluster power consumption (Figure 7.5). Therefore, in this regard, clusters 3 and 0 perform best in energy efficiency because they combine the highest amounts of high-quality grade scrap varieties (SS 300, SS 400, and SS Roll). Most of such lower-grade scraps are found in cluster 2, which is the least energy-efficient option (DIR, JAM, SAF Metal, and Tundish). According to the rules governing scrap categorizing, there are more contaminants in lower categories of scrap. Even during the manufacturing process, a slag layer made of these contaminants develops on top of the mixture of steel (Teo et al., 2016). The slag outlet in the EAF removes this layer, which takes time and increases the processing time and overall power consumption (Björkman & Samuelsson, 2014).

From a different angle, the amount of charged DRI and pig iron inside the heat and even the amount of power used by the EAF are positively correlated (Kirschen et al., 2021). It is seen in the cluster analysis, where the minuscule DRI portion in clusters 3 and 0 results in the most remarkable performance in terms of power consumption. On the other hand, clusters 2 and 1 combine with a substantial amount of DRI and Tundish to produce a high-power demanding method (Kirschen et al., 2011). Due to the oxide sludge in DRI, relative to temperatures of steel scrap of high or medium quality, the steel yield of DRI loaded is frequently lower. As a result, DRI and pig iron temperatures have a greater specific energy need than heat for liquefying steel scrap (Kirschen et al., 2021). In cluster 3, the two conditions

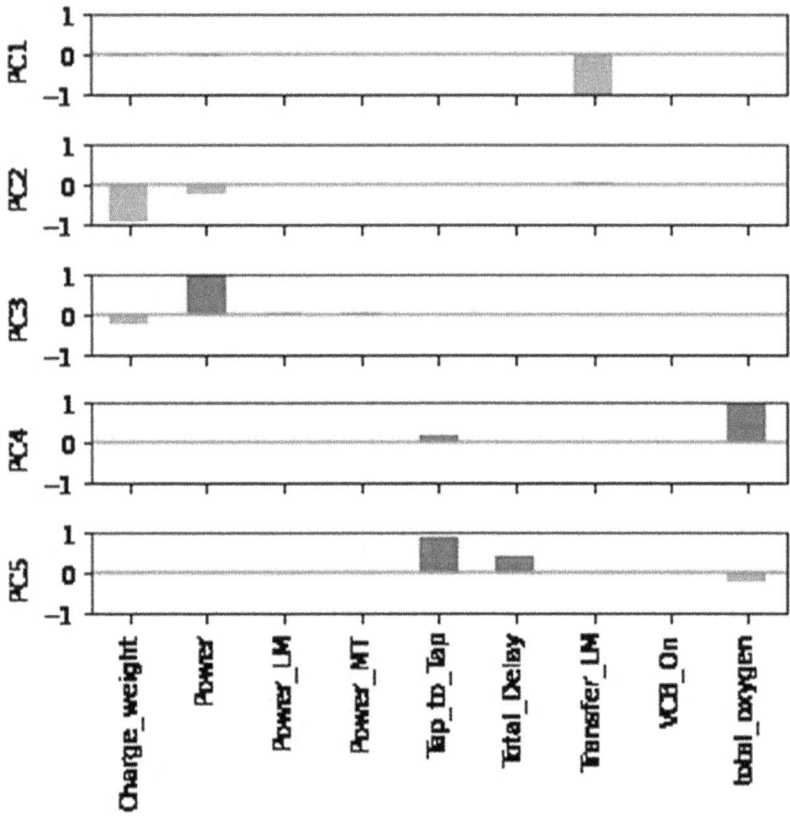

Figure 7.4 Principal component analysis.

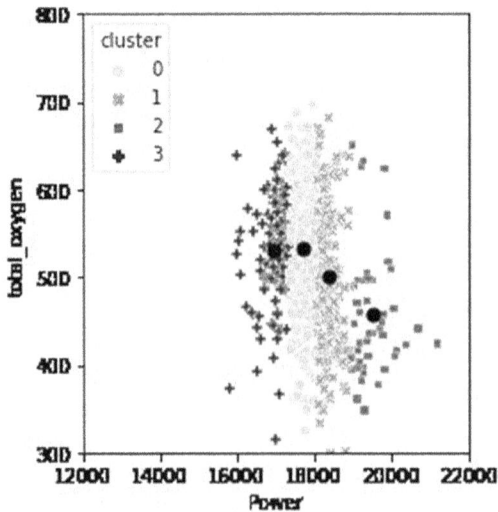

Figure 7.5 K-Means Clustering of heat data set.

(high-grade scraps such as SS 300, HC FE Cr, and a high proportion of DRI) when combined melt the input material to produce an intermediate power usage.

7.4.1 Managerial implications

The K-Means Clustering was used for the data analysis of a selected grade of steel in this study. It was found that liquifying of the steel scrap consumes a lot of energy in the EAF; from a different angle, the amount of charged DRI and pig iron inside the heat and even the amount of power used by the EAF are positively correlated. It can be also seen in the cluster analysis, where a minuscule DRI portion results in the most remarkable performance in terms of power consumption. The study of data analysis leads to the following implications, which are useful for the managers:

- Operations and maintenance managers will be able to improve the energy usage of an operation.
- It depends on the known input material by taking into account the individual behavior of production situations.
- An operations supervisor, for instance, would've been able to create a power production schedule and assign various materials to each heat to minimize the process's electricity use.
- The outcomes of cluster analysis can aid scrap purchase managers in fine-tuning their periodical purchase procedures at the support activity level.
- The everyday scrap inventory could accommodate one of the most power-efficient production clusters.

7.5 CONCLUSION LIMITATIONS AND FUTURE SCOPE

A data-driven statistical model is provided in this research to improve the power consumption performance of ecological manufacturing systems. The statistical model was used to analyze the power usage of an EAF in a stainless-steel production plant in India, a severe energy challenge. An advisory group presented the most affected parameters on the power consumption of EAF in this respect. Data was built from the logs of different heats for six months and a given grade. We gathered parameters with the help of industry experts, and K-Means Clustering techniques were constructed using those parameters. When these insights are combined with policy and operational methods, benefits like energy conservation, lower cost, and improved process regulation may be attained.

This study provides an integrated data-mining strategy for improving power consumption optimization in manufacturing systems, which it then implements in a real-world scenario in stainless-steel manufacturing. However, this study has some limitations, such as this whole study is subjected to a particular grade only; because of the EAF's complicated and chaotic operational nature, some production concerns should also be investigated in real-world manufacturing systems. As a result, future studies could look into the following three areas: (1) considering a broader range of impactful elements involved in the stainless-steel industry, like different grade specifications, recoveries, and certain other variables that impact power usage, (2) exploring production data over long durations and much more comprehensive and varied instances, and (3) applying the method to far more diverse data, including the composition of each raw material that is being input into the furnace.

REFERENCES

Argiolas, L., & Bacchetti, L. (2020). Q-SYM2 and Q-ASC: ABS scrap yard management, assisted scrap classification and digital integration into the steel process. *AISTech - Iron and Steel Technology Conference Proceedings* 1, 519–525. https://doi.org/10.33313/380/059

Bai, E. (2014). Minimizing energy cost in electric arc furnace steel making by optimal control designs. *Journal of Energy* 2014, 1–9. https://doi.org/10.1155/2014/620695

Bălan, R., Hancu, O., & Lupu, E. (2007). Modeling and adaptive control of an electric arc furnace. *IFAC Proceedings* 40(8), 163–168. https://doi.org/10.3182/20070709-3-RO-4910.00026

Bekker, J. G., Craig, I. K., & Pistorius, P. C. (1998). Modelling and specification for control of an EAF off-gas process. *IFAC Proceedings* 31(23), 55–60. https://doi.org/10.1016/S1474-6670(17)35856-1

Bekker, J. G., Craig, I. K., & Pistorius, P. C. (1999). Model predictive control of an EAF off-gas process. *IFAC Proceedings* 32(2), 7050–7055. https://doi.org/10.1016/S1474-6670(17)57203-1

Bekker, J. G., Craig, I. K., & Pistorius, P. C. (2000). Model predictive control of an electric arc furnace off-gas process. *Control Engineering Practice* 8(4), 445–455. https://doi.org/10.1016/S0967-0661(99)00163-X

Björkman, B., & Samuelsson, C. (2014). Recycling of steel. *Handbook of Recycling: State-of-the-Art for Practitioners, Analysts, and Scientists*, 65–83. https://doi.org/10.1016/B978-0-12-396459-5.00006-4

Boulet, B., Lalli, G., & Ajersch, M. (2003). Modeling and control of an electric arc furnace. *Proceedings of the American Control Conference* 4, 3060–3064. https://doi.org/10.1109/ACC.2003.1243998

Çamdalı, Ü., & Tunç, M. (2002). Modelling of electric energy consumption in the AC electric arc furnace. *International Journal of Energy Research*, 26(10), 935–947.

Carlsson, L. S., Samuelsson, P. B., & Jönsson, P. G. (2019). Predicting the electrical energy consumption of electric arc furnaces using statistical modeling. *Metals* 9(9), 959. https://doi.org/10.3390/MET9090959

Carlsson, L. S., Samuelsson, P. B., & Jönsson, P. G. (2020a). Interpretable ML —tools to interpret the predictions of a machine learning model predicting the electrical energy consumption of an electric arc furnace. *Steel Research International* 91(11). https://doi.org/10.1002/srin.202000053

Carlsson, L. S., Samuelsson, P. B., & Jönsson, P. G. (2020b). Modeling the effect of scrap on the electrical energy consumption of an electric arc furnace. *Processes* 8(9), 1044. https://doi.org/10.3390/PR8091044

Chang, G. W., Chen, C. I., & Liu, Y. J. (2010). A neural-network-based method of modeling electric arc furnace load for power engineering study. *IEEE Transactions on Power Systems* 25(1), 138–146. https://doi.org/10.1109/TPWRS.2009.2036711

Czapla, M., Karbowniczek, M., & Michaliszyn, A. (2008). The optimisation of electric energy consumption in the electric ARC furnace. *Archives of Metallurgy and Materials* 53(2), 559–565.

Gajic, D., Savic-Gajic, I., Savic, I., Georgieva, O., & di Gennaro, S. (2016). Modelling of electrical energy consumption in an electric arc furnace using artificial neural networks. *Energy* 108, 132–139. https://doi.org/10.1016/J.ENERGY.2015.07.068

Huang, X., & Han, M. (2008). Greedy kernel components acting on ANFIS to predict BOF steelmaking endpoint. *IFAC Proceedings* 41(2), 11007–11012. https://doi.org/10.3182/20080706-5-KR-1001.01864

Kaut, M. (2021). Scenario generation by selection from historical data. *Computational Management Science* 18(3), 411–429. https://doi.org/10.1007/S10287-021-00399-4/FIGURES/5

Kim, M. S., Choi, Y. J., Park, I. S., Kong, N., Lee, M., & Park, P. (2018). Sensitivity analysis on a neural network for analyzing the camber in hot rolling process. *Proceedings - 2018 IEEE Industrial Cyber-Physical Systems, ICPS 2018*, 212–216. https://doi.org/10.1109/ICPHYS.2018.8387661

Kirschen, M., Badr, K., & Pfeifer, H. (2011). Influence of direct reduced iron on the energy balance of the electric arc furnace in steel industry. *Energy* 36(10), 6146–6155. https://doi.org/10.1016/J.ENERGY.2011.07.050

Kirschen, M., Hay, T., & Echterhof, T. (2021). Process improvements for direct reduced iron melting in the electric arc furnace with emphasis on slag operation. *Processes 2021* 9(2), 402. https://doi.org/10.3390/PR9020402

Kordos, M., Blachnik, M., & Wieczorek, T. (2011). Temperature prediction in electric arc furnace with neural network tree. *Lecture Notes in Computer Science (Including Subseries Lecture Notes in Artificial Intelligence and Lecture Notes in Bioinformatics)* 6792 LNCS(PART 2), 71–78. https://doi.org/10.1007/978-3-642-21738-8_10/COVER/

Kovačič, M., Stopar, K., Vertnik, R., & Šarler, B. (2019). Comprehensive electric arc furnace electric energy consumption modeling: A pilot study. *Energies 2019* 12(11), 2142. https://doi.org/10.3390/EN12112142

Liu, J., Liu, X., & Le, B. T. (2019). Rolling force prediction of hot rolling based on GA-MELM. *Complexity* 2019. https://doi.org/10.1155/2019/3476521

Mesa Fernández, J. M., Cabal, V. Á., Montequin, V. R., & Balsera, J. V. (2008). Online estimation of electric arc furnace tap temperature by using fuzzy neural networks. *Engineering Applications of Artificial Intelligence* 21(7), 1001–1012. https://doi.org/10.1016/J.ENGAPPAI.2007.11.008

Mosavi, A., Salimi, M., Faizollahzadeh Ardabili, S., Rabczuk, T., Shamshirband, S., & Varkonyi-Koczy, A. R. (2019). State of the art of machine learning models in energy systems, a systematic review. *Energies*, 12(7), 1301.

Murua, M., Boto, F., Anglada, E., Cabero, J. M., & Fernandez, L. (2020). A slag prediction model in an electric arc furnace process for special steel production. *Procedia Manufacturing* 54, 178–183. https://doi.org/10.1016/j.promfg.2021.07.027

Olabi, A. G., Casalino, G., Benyounis, K. Y., & Hashmi, M. S. J. (2006). An ANN and Taguchi algorithms integrated approach to the optimization of CO2 laser welding.

Advances in Engineering Software 37(10), 643–648. https://doi.org/10.1016/J. ADVENGSOFT.2006.02.002

Oosthuizen, D. J., Craig, I. K., & Pistorius, P. C. (2004). Economic evaluation and design of an electric arc furnace controller based on economic objectives. *Control Engineering Practice* 12(3), 253–265. https://doi.org/10.1016/ S0967-0661(03)00078-9

Richards, A., & How, J. (2005). Mixed-integer programming for control. *Proceedings of the American Control Conference* 4, 2676–2683. https://doi.org/10.1109/ ACC.2005.1470372

Saboohi, Y., Fathi, A., Skrjanc, I., & Logar, V. (2019). Optimization of the electric arc furnace process. *IEEE Transactions on Industrial Electronics* 66(10), 8030–8039. https://doi.org/10.1109/TIE.2018.2883247

Saparrat, M., Monti, F., & Ibarra, J. (2020). AI Application to melting temperature prediction in an electric arc furnace. *AISTech - Iron and Steel Technology Conference Proceedings* 1, 526–532. https://doi.org/10.33313/380/060

Shanno, D. F. (1970). Conditioning of quasi-Newton methods for function minimization. *Mathematics of Computation* 24(111), 647–656. https://doi.org/10.1090/ S0025-5718-1970-0274029-X

Shyamal, S., & Swartz, C. L. E. (2018). Real-time dynamic optimization-based advisory system for electric arc furnace operation. *Industrial and Engineering Chemistry Research* 57(39), 13177–13190. https://doi.org/10.1021/ACS.IECR. 8B02542/ASSET/IMAGES/MEDIUM/IE-2018-02542W_0007.GIF

Shyamal, S., & Swartz, C. L. E. (2019). Real-time energy management for electric arc furnace operation. *Journal of Process Control* 74, 50–62. https://doi. org/10.1016/J.JPROCONT.2018.03.002

Teo, P., Seman, A. A., Basu, P., & Sharif, N. M. (2016). Characterization of EAF steel slag waste: The potential green resource for ceramic tile production. *Procedia Chemistry* 19, 842–846. https://doi.org/10.1016/J.PROCHE.2016.03.111

Todshki, N. E., & Ranjbaraki, A. (2016). The impact of major macroeconomic variables on Iran's steel import and export. *Procedia Economics and Finance* 36, 390–398. https://doi.org/10.1016/S2212-5671(16)30051-X

Tomažič, S., Andonovski, G., Škrjanc, I., & Logar, V. (2022). Data-driven modelling and optimization of energy consumption in EAF. *Metals* 12(5). https://doi. org/10.3390/met12050816

Toulouevski, Y. N., & Zinurov, I. Y. (2013). Modern steelmaking in electric arc furnaces: History and development. *Innovation in Electric Arc Furnaces* 1–24. https://doi.org/10.1007/978-3-642-36273-6_1

Urbański, T. S. (1974). Extractive separation of a lanthanum tracer from preparations on a calcium carrier obtained from steel samples. *Journal of Radioanalytical Chemistry* 22(1–2), 13–19. https://doi.org/10.1007/BF02518088

Wang, F., Jin, Z., & Zhu, Z. (2005). Modeling and prediction of electric arc furnace based on neural network and chaos theory. *Lecture Notes in Computer Science* 3498(III), 819–826. https://doi.org/10.1007/11427469_130/COVER/

Zhou, Y., & Forbes, J. F. (2003). Determining controller benefits via probabilistic optimization. *International Journal of Adaptive Control and Signal Processing* 17(7–9), 553–568. https://doi.org/10.1002/ACS.765

Chapter 8

PID-based ANN control of dynamic systems

A. Kharola

Graphic Era Deemed to be University, Dehradun, India

CONTENTS

8.1 INTRODUCTION

The inverted pendulum systems are inherently unstable, underactuated mechanical systems which are difficult to control (Bayram & Kara, 2020). These systems are widely used for verification of various control laws and emerging approaches (Timmermann et al., 2011). There exist different configurations of inverted pendulum systems based on their field of application. Inverted double pendulum is one of such complex configuration which includes two pendulums placed one over another and pivoted to a moving cart (Votrina et al., 2021). The two links attached to the cart make the system more difficult to control. An actuation force is needed to be produced at the joints of cart and pendulum assembly, which helps in keeping the pendulums at upright position. Many researchers have successfully controlled these systems adopting various control techniques. Henmi et al. (2004) proposed energy-based control of serial- and parallel-type double inverted pendulum. A robust sliding mode controller (SMC) has also been developed which enhances the robustness of the proposed system.

Punlabpho and Jearsiripongkul (2015) developed a robust controller based on H-infinity control for stabilization of double inverted pendulum system. The results showed that the controller was able to stabilize the proposed system in an upright position even when subjected to external disturbances. Guo and Xiong (2018) combined fuzzy approximation and integer backstepping technique to develop a novel time-delay assumption-independent state feed-back control scheme for control of double inverted pendulum. Simulation results confirmed the effectiveness of the proposed controller. Jaafar et al. (2018) proposed a proportional-integral-derivative (PID) control of over-head crane modeled as a double pendulum system. The controller was able to stabilize the complete system without needing payload motion feedback signal. The parameters of the PID controller were tuned using particle swarm optimization (PSO) technique. The results showed superiority of the proposed controller compared to previously developed PSO-tuned PID controller.

An input-output inversion technique for open-loop control of an over-head crane modeled as a double pendulum was proposed by Giacomelli et al. (2018). The parametric trajectory of the system has been analyzed mathematically and resulted in reduction of residual oscillations. The results confirm that the proposed technique can be incorporated with industrial drives after neglecting post-actuations. Coxe (2019) developed a proportional derivative (PD) controller for control of an inverted double pendulum. The responses of PD controller were used for training of artificial neural networks (ANNs). The results showed better response of ANN controller in terms of settling time and overall stability. Mehedi et al. (2019) developed a fractional order controller based on Bode's ideal transfer function for double inverted rotary pendulum. The authors further compared proposed controller with an integer order controller based on state space approach. The results showed better performance of fractional order controller showing a smooth voltage control and more robustness to external disturbances. Kim and Lee (2021) proposed a PID controller incorporated with switching action for control of non-linear systems. The proposed controller exhibits simplicity of a PID controller and robustness of an SMC. Experimental results proved the authenticity of the proposed controller.

Sanjeewa and Parnichkun (2022) developed a linear quadratic regulator (LQR)-based SMC for control of rotary double inverted pendulum. Euler-Lagrange formulation was considered for developing a dynamic model of the system. The results indicate better performance of the proposed controller compared to conventional SMC controller. In this chapter, an ANN controller has been designed based on the PID controller. A mathematical model of the system has been developed and simulated in MATLAB®/Simulink. The system configuration includes two rigid pendulums of masses ($m_1 = m_2 = 0.5$ kg) and lengths ($l_1 = l_2 = 0.1$ meter) inclined at an angle θ_1 and θ_2, respectively, with the vertical, as shown in Figure 8.1. The lower pendulum is attached to a cart or carriage of mass (M=1kg). The carriage can move in linear direction under the influence of control force (F) and acceleration due to gravity (g=9.81m/s²). The moment of inertia (I) for the pendulums is 0.006 Kgm², and friction coefficient (b) acting between cart and surface is 0.1 Ns/m².

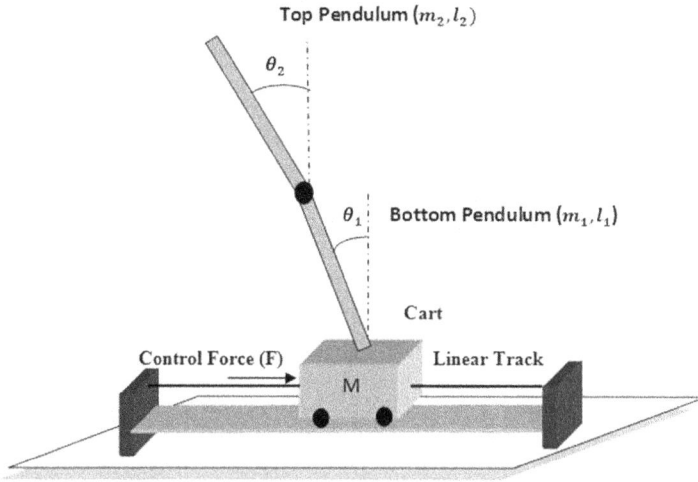

Figure 8.1 Two-stage cart and pendulum system.

8.2 MATHEMATICAL MODELING OF INVERTED DOUBLE PENDULUM

Newton's second law has been incorporated for deriving non-linear differential equations of the proposed system (Myers et al., 2020). These equations give expression for linear acceleration of carriage ($\ddot{x}$) and angular accelerations of pendulums ($\ddot{\theta}$). The forces acting on carriage sub-system is shown in Figure 8.2. In the figure, N_1 and P_1 indicate the interaction forces between lower pendulum and carriage, respectively, in horizontal and vertical directions. The frictional force acting on the carriage is given by $b\dot{x}$.

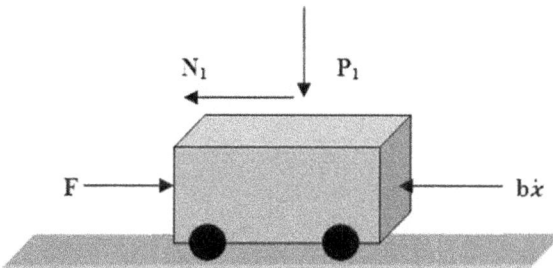

Figure 8.2 Forces acting on carriage sub-system.

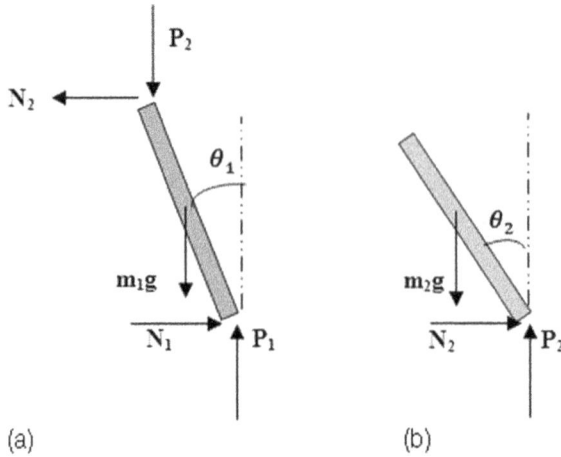

Figure 8.3 Forces acting on (a) lower and (b) upper pendulum.

After equating forces in horizontal direction, the following expression for linear acceleration of the carriage is obtained, as shown in Equation (8.1):

$$\ddot{x} = \frac{1}{M}\left(F - N_1 - b\dot{x}\right) \tag{8.1}$$

Here, $\dot{x}$ and $\ddot{x}$ represent the linear velocity and linear acceleration of the carriage, respectively. The forces acting on lower and upper pendulum are shown in Figure 8.3. In the figure, N_2 and P_2 indicate the interaction forces acting between the pendulums in, respectively, horizontal and vertical directions. Additionally, m_1g and m_2g indicate the gravitational forces acting on, respectively, the lower and upper pendulums.

The angular acceleration ($\ddot{\theta_1}$) of the lower pendulum is given by Equation (8.2).

$$\ddot{\theta_1} = \frac{1}{I_1}\left(N_1l_1\cos\theta_1 + P_1l_1\sin\theta_1 - b_1\dot{\theta_1} + N_2l_1\cos\theta_1 + P_2l_1\sin\theta_1\right) \tag{8.2}$$

Here, $\dot{\theta_1}$ and I_1 are angular velocity and moment of inertia for lower pendulum, respectively.

The angular acceleration ($\ddot{\theta_2}$) of the upper pendulum is given by Equation (8.3).

$$\ddot{\theta_2} = \frac{1}{I_2}\left(N_2l_2\cos\theta_2 + P_2l_2\sin\theta_2 - b_2\dot{\theta_2}\right) \tag{8.3}$$

Here, $\dot{\theta_2}$ and I_2 are angular velocity and moment of inertia for upper pendulum, respectively. Further, a simulink model of the proposed system has been developed considering above equations, as shown in Figure 8.4.

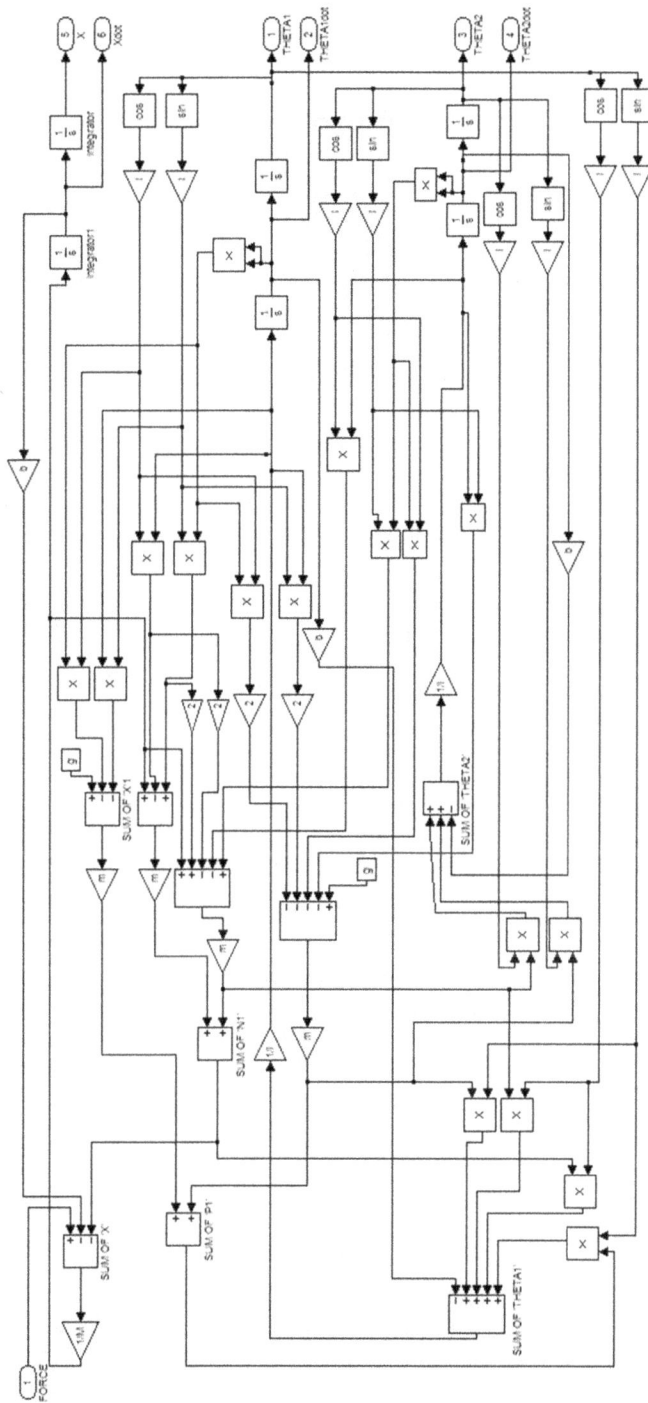

Figure 8.4 Simulink model of inverted double pendulum system.

8.3 PID-BASED ANN CONTROL OF INVERTED DOUBLE PENDULUM SYSTEM

In this study, the PID controller has been tuned using trial and error method (Barreiros et al., 2021). PID gains obtained after tuning for different sub-systems are given in Table 8.1.

The results obtained after simulation of the PID controller were stored in MATLAB® workspace and used for training of neural controller. The data samples were arbitrary divided into training, validation, and testing samples, as shown in Table 8.2. These samples were further used for the designing of three neural controllers to control each sub-system separately.

A neural controller has been developed with two inputs and one output architecture having 20 neurons in hidden layer and 1 neuron in output layer, as shown in Figure 8.5.

Neural controller has been trained using Levenberg-Marquardt Back-Propagation algorithm (Al-Mayyahi et al., 2015). The parameters have been optimized using stochastic gradient descent optimization algorithm. The regression results for training, validation, and testing samples clearly indicate a strong correlation between outputs and targets as shown in Figure 8.6. In addition, minimum MSE values for all the controllers have been obtained, as shown in Table 8.3.

Table 8.1 PID gains obtained for different sub-systems

Sub-system	Integral gain (K$_i$)	Proportional gain (K$_p$)	Differential gain (K$_d$)
Upper pendulum	1	1	0
Lower pendulum	1	1	0
Carriage	1	1	0

Table 8.2 Data samples for training of neural controllers

Type of sample	Percentage share (%)	Sample size
Training	70	111
Validation	15	24
Testing	15	24

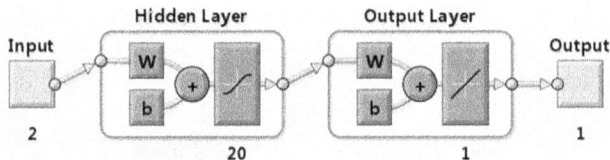

Figure 8.5 Neural controller architecture.

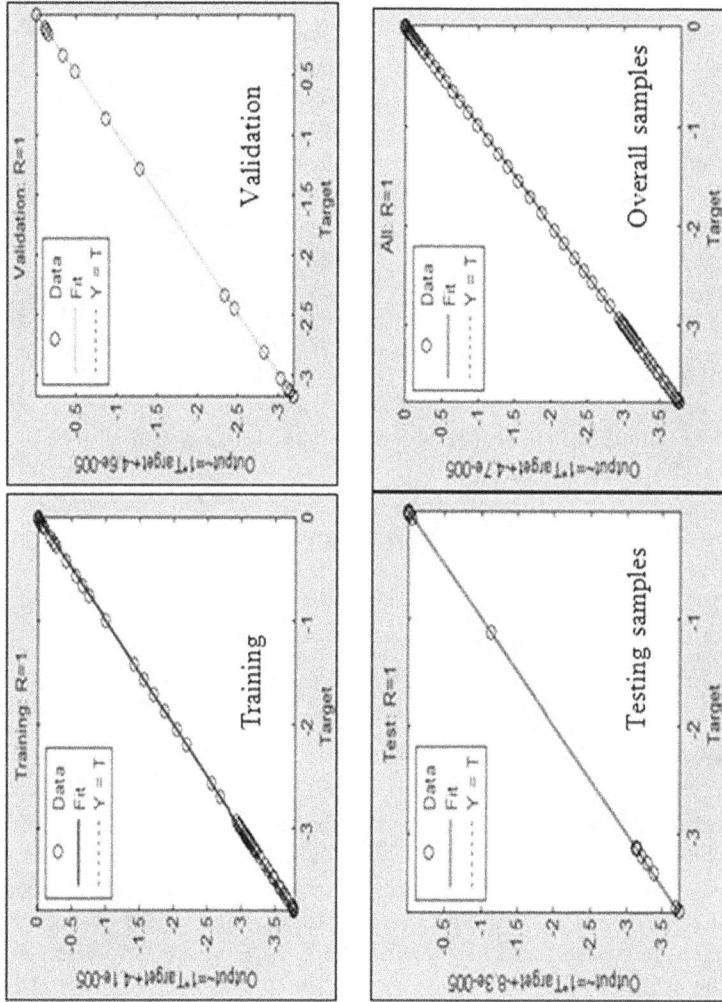

Figure 8.6 Regression results for different samples of top pendulum controller.

Table 8.3 MSE and regression values obtained for neural controllers

Top pendulum controller		
Type of sample	MSE	Regression
Training	1.08e-9	1
Validation	1.44e-8	1
Testing	2.43e-8	1
Bottom pendulum controller		
Training	9.53e-10	1
Validation	1.03e-6	1
Testing	2.58e-6	1
Cart controller		
Training	6.76e-9	1
Validation	4.57e-9	1
Testing	8.98e-8	1

8.4 SIMULATION AND RESULTS COMPARISON

The PID-based neural controller has been simulated using a simulink sub-system, as shown in Figure 8.7.

The simulation responses obtained for inverted double pendulum system using above controllers are shown with the help Figures 8.8–8.13. The time considered for simulation was 5 sec.

A comparison of simulation results for PID and neural controllers is shown in Table 8.4. The parameters considered for the comparison include settling time, overshoot ranges, and steady-state error.

8.5 CONCLUSION

This chapter presents PID-based neural control of inverted double pendulum system. The result of the PID controller has been used for training parameters of neural controllers. The neural controller were trained till optimal values of regression and MSE had been obtained. The number of neurons in hidden layer has been increased till minimum MSE values and unit regression value were obtained for all the controllers. The performance of both controllers was monitored in terms of settling time, overshoot response, and steady-state error. The results indicate better performance of neural controller specifically in terms of settling time. The neural controller has been able to control upper pendulum angle, lower pendulum angle, and carriage position within 2.7 sec, 2.8 sec, and 3.2 sec, respectively. However, the times taken by PID for controlling the same attributes were 3.0 sec, 3.1 sec and 3.3 sec, respectively. Similarly, the time taken by neural controller to stabilize

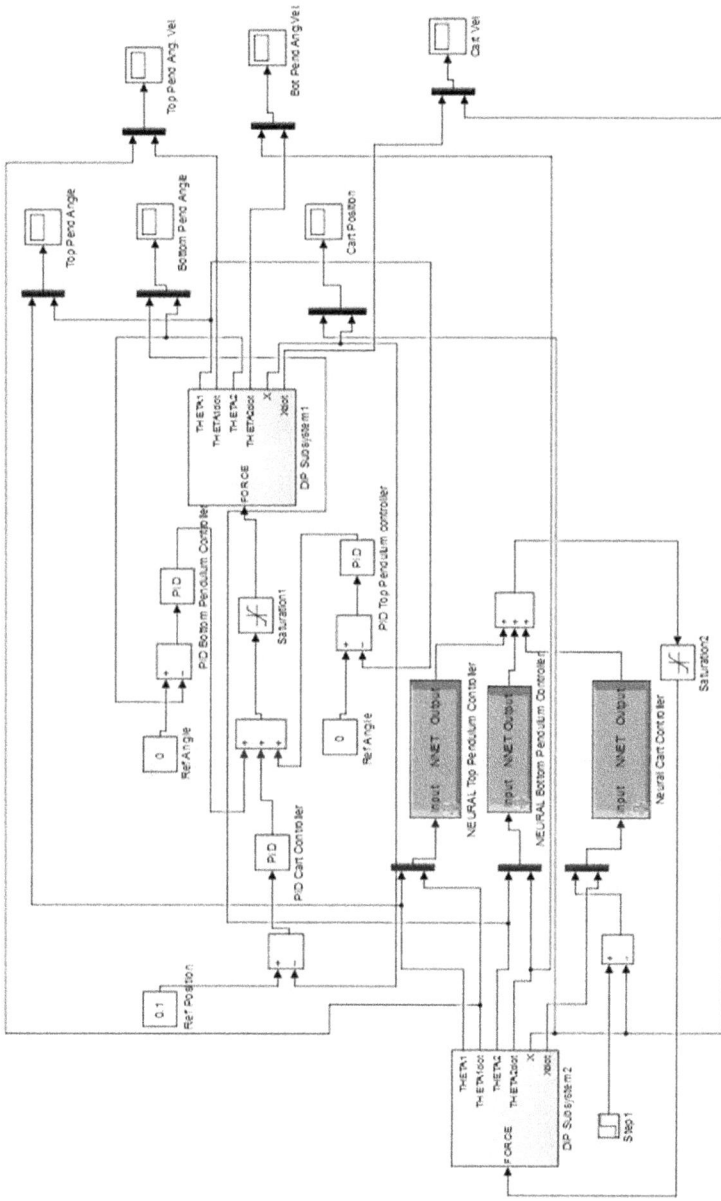

Figure 8.7 Simulink sub-system for comparing performance of PID-based neural controller.

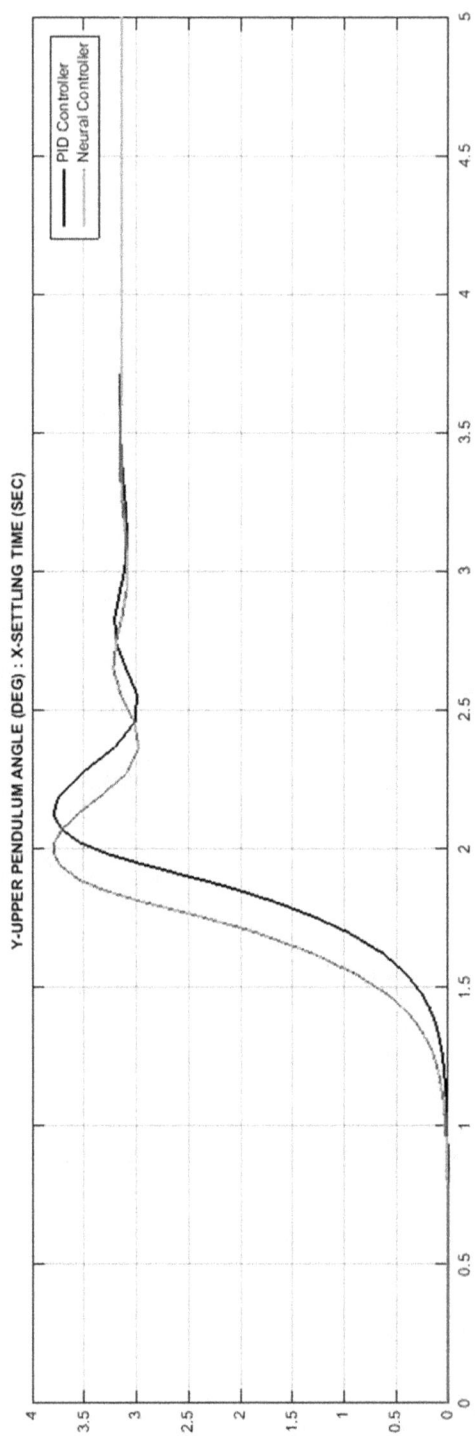

Figure 8.8 Responses obtained for 'Upper pendulum angle'.

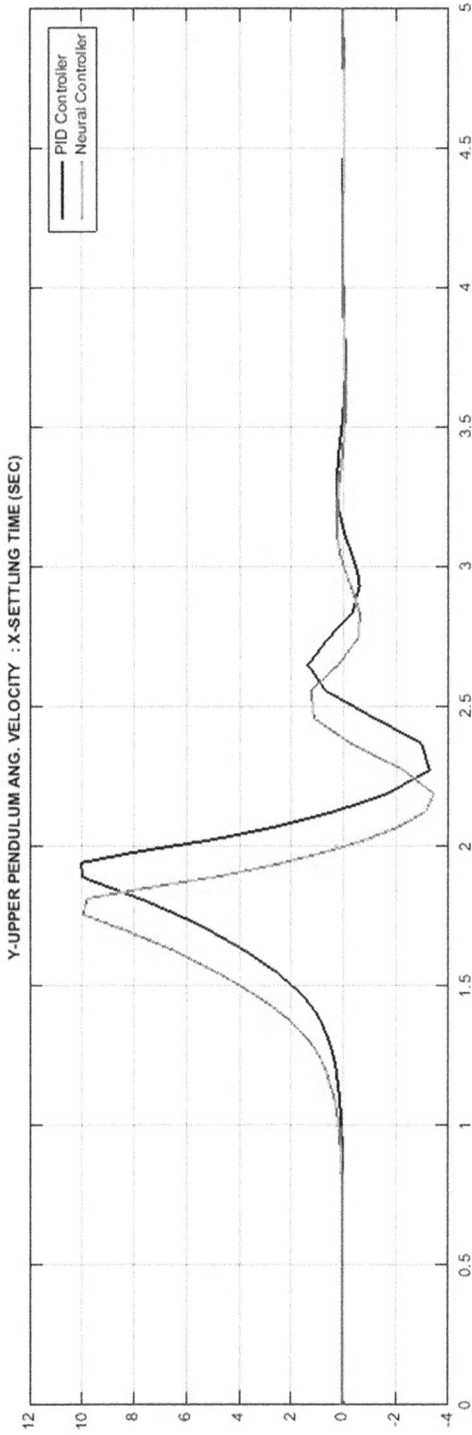

Figure 8.9 Responses obtained for 'Upper pendulum angular velocity'.

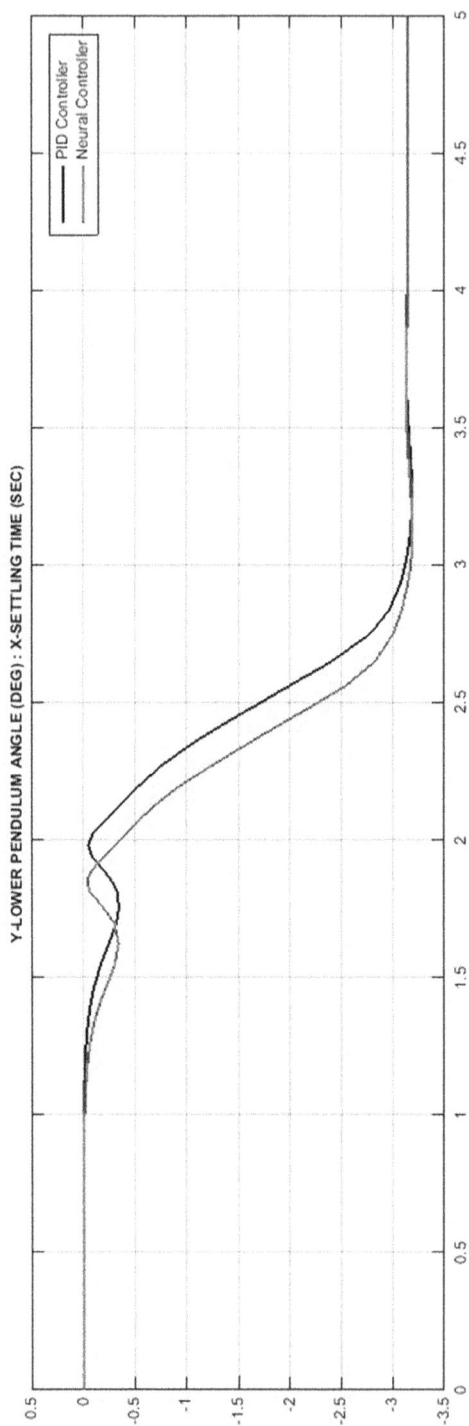

Figure 8.10 Responses obtained for 'Lower pendulum angle'.

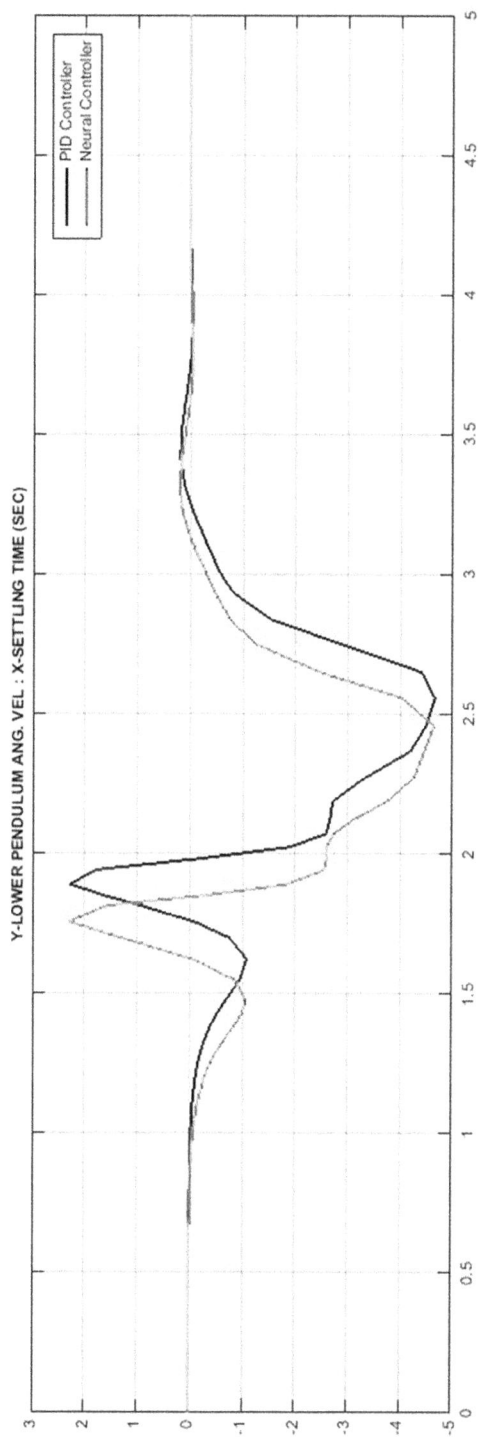

Figure 8.11 Responses obtained for 'Lower pendulum angular velocity'.

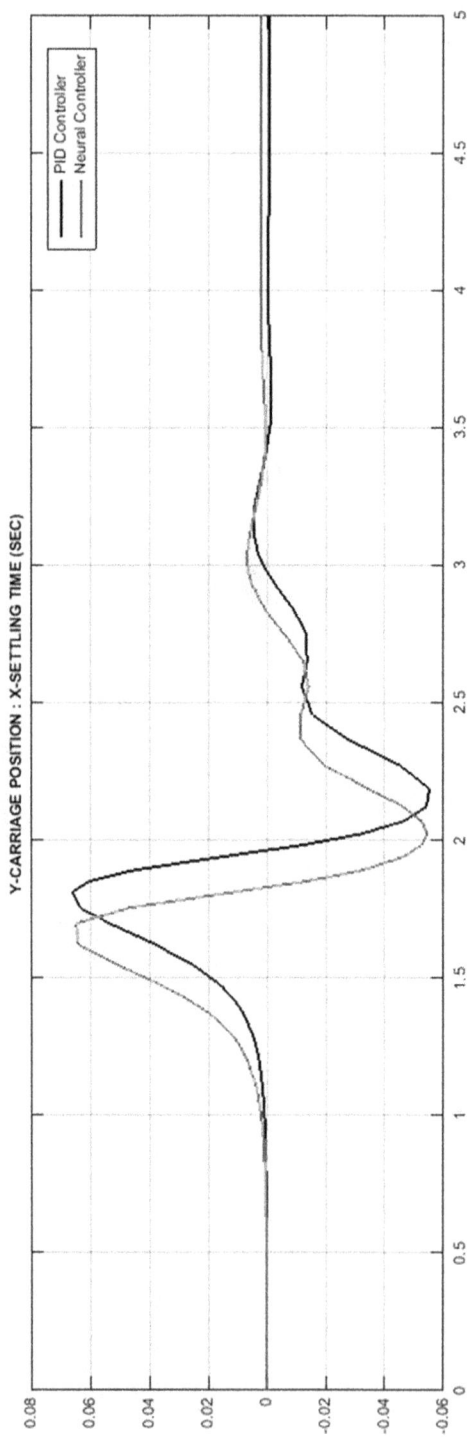

Figure 8.12 Responses obtained for 'Carriage position'.

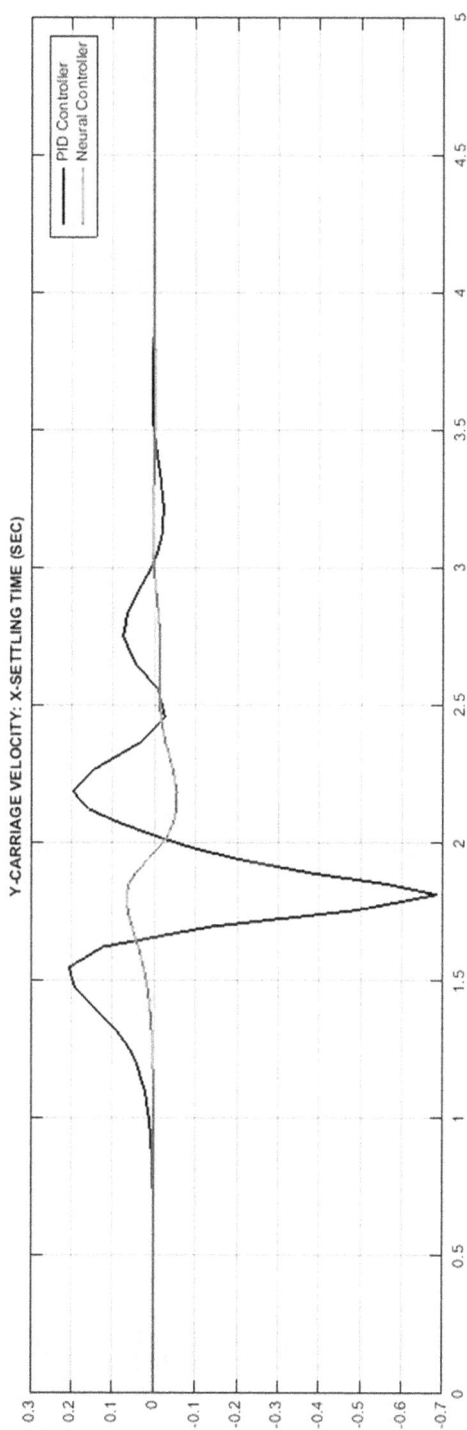

Figure 8.13 Responses obtained for 'Carriage velocity'.

Table 8.4 Simulation results obtained using PID and neural controllers

Controller	Settling time (seconds)	Overshoot ranges (degree)	Steady-state error
Upper pendulum angle			
PID	3.0 sec	3.7°	0
Neural	2.7 sec	3.7°	0
Upper pendulum angular velocity			
PID	3.3 sec	10.2° to −3.4°	0
Neural	3.1 sec	10.0° to −3.5°	0
Lower pendulum angle			
PID	3.1 sec	−3.2°	0
Neural	2.8 sec	−3.2°	0
Lower pendulum angular velocity			
PID	3.8 sec	2.2° to −4.6°	0
Neural	3.6 sec	2.2° to −4.6°	0
Cart position			
PID	3.3 sec	0.07° to −0.05°	0
Neural	3.2 sec	0.07° to −0.05°	0
Cart velocity			
PID	3.5 sec	0.2° to −0.68°	0°
Neural	2.8 sec	0.08° to −0.06°	0

upper pendulum angular velocity, lower pendulum angular velocity, and cart velocity were 3.1 sec, 3.6 sec, and 2.8 sec, respectively. However, the time taken by the PID controller for stabilizing the same attributes were 3.3 sec, 3.8 sec, and 3.5 sec, respectively. Furthermore, it is also observed that both the controllers show good performance toward overshoot ranges and steady-state error.

REFERENCES

Al-Mayyahi, A., Wang, W., and Birch, P. 2015. Levenberg-Marquardt optimized neural networks for trajectory tracking of autonomous ground vehicles. *International Journal of Mechatronics and Automation* 5(2/3):140–153, doi: 10.1504/IJMA.2015.075960.

Barreiros, B.F., Trierweiler, J.O., and Farenzena, M. 2021. Reliable and straightforward PID tuning rules for highly underdamped systems. *Brazilian Journal of Chemical Engineering* 38:731–745, doi: 10.1007/s43153-021-00127-0.

Bayram, A., and Kara, F. 2020. Design and control of spatial inverted pendulum with two degrees of freedom. *Journal of the Brazilian Society of Mechanical Sciences and Engineering* 42:1–14, doi: 10.1007/s40430-020-02580-3.

Coxe, A. 2019. Neural control model for an inverted double pendulum. *Complex Systems* 28(2):239–249, doi: 10.25088/ComplexSystems.28.2.239

Giacomelli, M., Padula, F., Simoni, L., and Visioli, A. 2018. Simplified input-output inversion control of a double pendulum overhead crane for residual oscillation reduction. *Mechatronics* 56:37–47, doi: 10.1016/j.mechatronics.2018.10.002.

Guo, T., and Xiong, J. 2018. A new global fuzzy fault-tolerant control for a double inverted pendulum based on time-delay replacement. *Neural Computing and Applications* 29(9):467–476, doi: 10.1007/s00521-016-2576-1.

Henmi, T., Deng, M., Inoue, A., Ueki, N., and Hirashima, Y. 2004. Energy-based control of a double inverted pendulum. *IFAC Proceedings Volumes* 37(12):173–178, doi: 10.1016/S1474-6670(17)31463-5.

Jaafar, H.I., Mohamed, Z., Md. Subha, N.A., Husain, A.R., Ismail, F.S., Ramli, L., Tokhi, M.O., and Shamsudin, M.A. 2018. Efficient control of a nonlinear double-pendulum overhead crane with sensorless payload motion using an improved PSO-tuned PID controller. *Journal of Vibration and Control* 25(4):907–921, doi: 10.1177/1077546318804319.

Kim, M.H., and Lee, S.U. 2021. PID with a switching action controller for nonlinear systems of second order controller canonical form. *International Journal of Control, Automation and Systems* 19:2343–2356, doi: 10.1007/s12555-020-0346-4.

Mehedi, I.M., Al-Saggaf, U.M., Mansouri, R., and Bettayeb, M. 2019. Stabilization of a double inverted rotary pendulum through fractional order integral control scheme. *International Journal of Advanced Robotic Systems* 16(4):1–9, doi: 10.1177/1729881419846741.

Myers, A.D., Templeman, J.R., Petrushenko, D., and Khasawneh, F.A. 2020. Low-cost double pendulum for high-quality data collection with open-source video tracking and analysis. *HardwareX* 8:1–23, doi: 10.1016/j.ohx.2020.e00138.

Punlabpho, S., and Jearsiripongkul, T. 2015. Control system for double inverted pendulum on a cart by H-infinity controller. *International Review of Automatic control* 8(4):300–306, doi: 10.15866/ireaco.v8i4.6968.

Sanjeewa, S.D.A., and Parnichkun, M. 2022. Control of rotary double inverted pendulum system using LQR sliding surface based sliding mode controller. *Journal of Control and Decision* 9(1):89–101, doi:10.1080/23307706.2021.1914758.

Timmermann, J., Khatab, S., Ober-Blobaum, S., and Trachtler, A. 2011. Discrete mechanics and optimal control and its application to a double pendulum on a cart. *IFAC Proceedings Volumes* 44(1):10199–10206, doi: 10.3182/20110828-6-IT-1002.01985.

Votrina, O.A., Meleshkin, K.N., and Frantsuzova, G.A. 2021. On the problem of synthesis of the controller based on sliding modes for a model object in the shape of a double inverted pendulum on a cart. *Optoelectronics, Instrumentation and Data Processing* 57:356–362, doi: 10.3103/S8756699021040129.

Chapter 9

Fatigue damage prognosis of offshore piping

A. Keprate
Oslo Metropolitan University, Oslo, Norway
TechnipFMC, Lysaker, Norway

N. Bagalkot
Oslo Metropolitan University, Oslo, Norway

CONTENTS

9.1 INTRODUCTION

The oil and gas being produced subsea are transported via pipelines and risers to the process facilities on an offshore platform, which contains both the process and the utility piping (UP). The distinction between the two piping systems is in terms of the fluid they carry. For example, process piping (PP) carries flammable fluids like oil, gas, condensate, glycol, etc., while the UP transports fluids (like cooling water, steam, etc.) necessary for supporting the main process operations. Although, both piping types undergo degradation since their installation, the degradation rate of PP is much higher than UP, primarily because of corrosive nature of hydrocarbons and high operating pressure and higher mechanical vibrations. At the same time, a hydrocarbon leak (HCL) from PP is more disastrous than that from UP. Figure 9.1 shows different equipments (of which PP tops the list) that are responsible for a major HCL on offshore platforms in the United Kingdom Continental Shelf

DOI: 10.1201/9781003434849-9

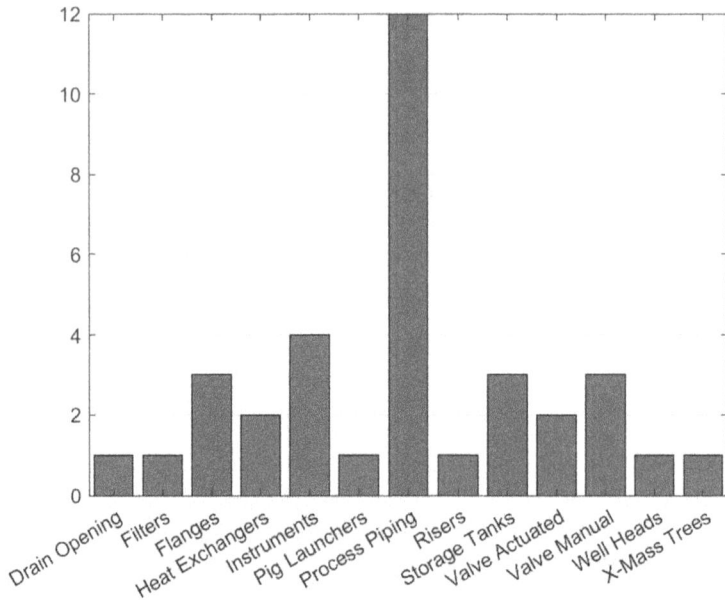

Figure 9.1 Equipment types responsible for major HCR in UKCS. (Adopted from HSE-RR672, 2008.)

(UKCS). Consequently, if operators want to preclude the HCL on offshore platform, they must perform pro-active maintenance of PP.

It is an industry practice to distinguish PP based upon its size. For example, PP, which has a nominal diameter of less than 50 mm, is called "small-bore piping" (SBP), while the PP having nominal diameter larger than 50 mm is termed "mainline piping" (MP). SBP generally branches off the MP with the purpose of acting either as an instrumentation port or a relief line or a drain vent. The connection between the SBP and MP is called a small-bore connection (SBC; as shown in Figure 9.2), and it can have different types of fittings such weldolet, socketolet, etc. It is at SBC, where the HCL generally emanates from, hence SBCs are the most critical fittings for the PP.

The root cause for HCL from PP is due to degradation mechanisms such as vibration-induced fatigue (VIF), acoustic-induced fatigue (AIF), pitting corrosion, corrosion under insulation, erosion, etc. A research study performed by DNV (results shown in Figure 9.3) clearly highlights that among the aforesaid degradation mechanisms, high-cycle fatigue (HCF) is the prominent cause of PP failure in petroleum and maritime sector. Hence, in this chapter we shall focus our discussion on fatigue failure of PP only.

9.2 UNDERSTANDING PIPING FATIGUE

In a PP system, cyclic loads originate due to the vibration of the MP carrying the hydrocarbons, eventually leading to the magnified vibrations in SBP and

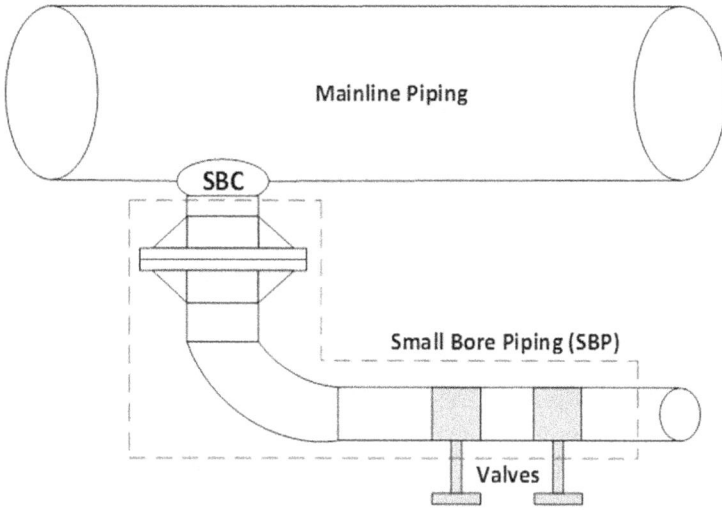

Figure 9.2 Diagrammatic representation of MP, SBP, and SBC. (Adapted from Harper, 2014.)

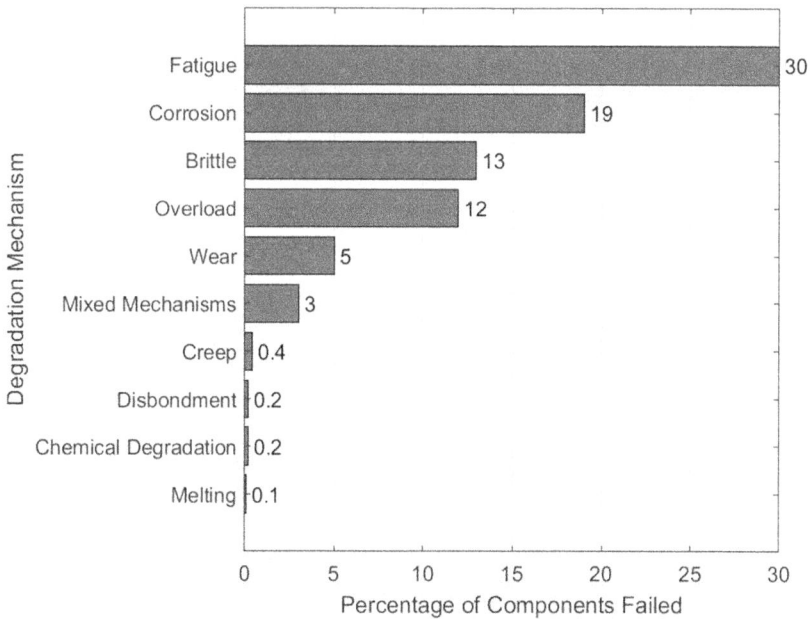

Figure 9.3 Dominant degradation mechanisms for process piping. (Adopted from DNV, 2020.)

ultimately causing the fatigue failure of the SBCs leading to HCL. Hence, from a process safety perspective it is necessary to comprehend the chain of actions causing the fatigue failure of the SBCs. Figure 9.4 depicts the failure chain due to the VIF.

It can be seen from Figure 9.4 that the excitation forces responsible for vibrating the MP are classified as steady state or transient. The former occurs during the normal operating conditions and emanates due to phenomenon such as flow-induced turbulence, flow-induced pulsations, pulsations from pumps/compressors, etc. (EI Guidelines, 2007). On the contrary, the latter originate during events such as changing operating conditions, additional units coming online, normal start-up and shutdown, emergency shutdown, valve operation, etc. (HSE-OTR-028, 2002). In order to mitigate the likelihood of the hydrocarbon leakage from the PP on the offshore platforms, it is necessary to deeply understand the excitation mechanisms depicted in Figure 9.4. The taxonomy of the fatigue originating in PP is illustrated in Figure 9.5. PP, especially near rotary equipment, is subjected to HCF, and VIF is the major reason for its failure. Although piping vibration is seldomly detected visually, nevertheless expert knowledge (from standards such as EI

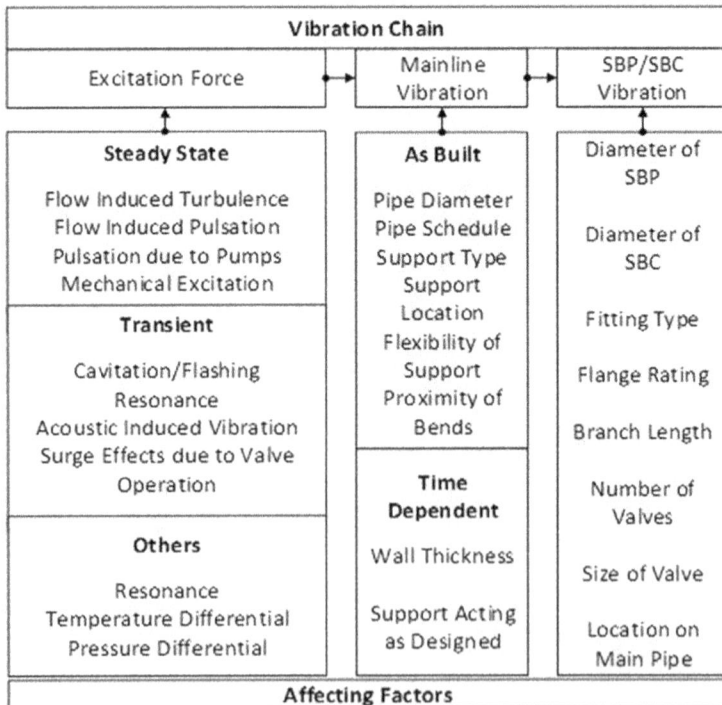

Vibration Chain		
Excitation Force	Mainline Vibration	SBP/SBC Vibration
Steady State Flow Induced Turbulence Flow Induced Pulsation Pulsation due to Pumps Mechanical Excitation	**As Built** Pipe Diameter Pipe Schedule Support Type Support Location Flexibility of Support Proximity of Bends	Diameter of SBP Diameter of SBC Fitting Type Flange Rating Branch Length
Transient Cavitation/Flashing Resonance Acoustic Induced Vibration Surge Effects due to Valve Operation		
	Time Dependent Wall Thickness Support Acting as Designed	Number of Valves Size of Valve Location on Main Pipe
Others Resonance Temperature Differential Pressure Differential		
Affecting Factors		

Figure 9.4 Vibration-induced fatigue failure chain. (Adapted from Swindell, 2003.)

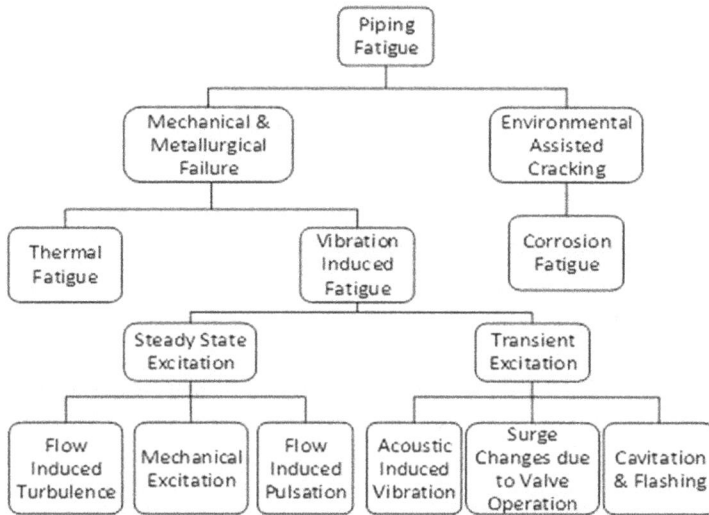

Figure 9.5 Taxonomy of the piping fatigue.

Guidelines and from practicing engineers) coupled with condition monitoring can help to frame effective inspection and maintenance (I&M) strategies for preventing unwanted piping breakdown due to fatigue.

9.3 FATIGUE DAMAGE PROGNOSIS

9.3.1 General

In the recent years, development of virtual sensors, coupled with cheap data storage, and advancement in AI, has enabled condition-based maintenance (CBM) as the most promising maintenance strategy for TPT. Prognostics, which deals with estimating the remnant useful life (RUL) of TPT is one of the vital steps in CBM. The main rationale behind prognostic models is their ability to estimate the future health of the piping system and thus generate warnings if the system is prone to failure. This ensures timely I&M of the piping system before these assets lead to an HCL, thus augmenting safety and mitigating downtime [(Chatterjee and Keprate, 2021)]. A spate amount of modeling approaches have been employed by the research community to perform fatigue damage prognostics of PP [(Keprate & Ratnayake, 2016)]. However, the two approaches that stand out are physics-based approaches and data-driven approaches. In particular for deteriorating due to fatigue, the former modeling approach is used to predict RUL, due to large availability of such models (such as Paris law and Elbers' model). Hence, from hereon we shall limit our discussion to physics-based crack growth models.

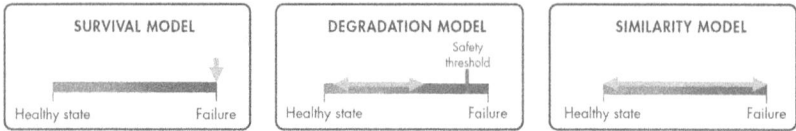

Figure 9.6 RUL estimator models. (Adapted from reference MATLAB®, 2022.)

9.3.2 Remaining useful life prediction

Common models employed for RUL estimation are survival model, degradation model, and similarity models, as shown in Figure 9.6. The choice of the model that shall be used for analysis is primarily dependent upon the amount of information at hand. Survival models are used when failure data of the equipment is available, while the degradation model finds its use when the data of healthy as well as failure state along with the safety threshold is available. In order to estimate the remaining fatigue life of SBPP, the degradation model (Paris law), coupled with uncertainty quantification and propagation, is used.

By performing integration of the Paris law, the RUL of the piping component can be estimated (Keprate et al., 2017) using the following equation

$$N = \frac{a_N^{1-\frac{m}{2}} - a_i^{1-\frac{m}{2}}}{C\left(1-\frac{m}{2}\right)\left(\Delta\sigma Y\sqrt{\pi}\right)^m}$$

On the one hand, while making a deterministic RUL prediction, all the parameters are treated as constants, on the other hand, for probabilistic RUL prediction, the variables are considered to be random as shown in Figure 9.7. The next step to uncertain quantification is to propagate it via Monte Carlo Simulation (MCS) to predict the confidence interval of the RUL (Keprate and Ratnayake, 2015, Engel et al., 2000). The next section deals with the deterministic and probabilistic RUL assessment of SBPP.

9.4 CASE STUDY

9.4.1 Background

The chosen material for SBPP for this case study is A106 GradeB, having a specific minimum yield strength of 240 MPa. For the purpose of illustration, it is assumed that a semi-elliptical surface crack has formed at the weld toe of the SBC. In order to capture the stress acting on the pipe walls due to the flowing fluid, a fluid structure interaction (FSI) analysis comprising of computational fluid dynamics (CFD) and structural finite element analysis (FEA) is performed, which is discussed next.

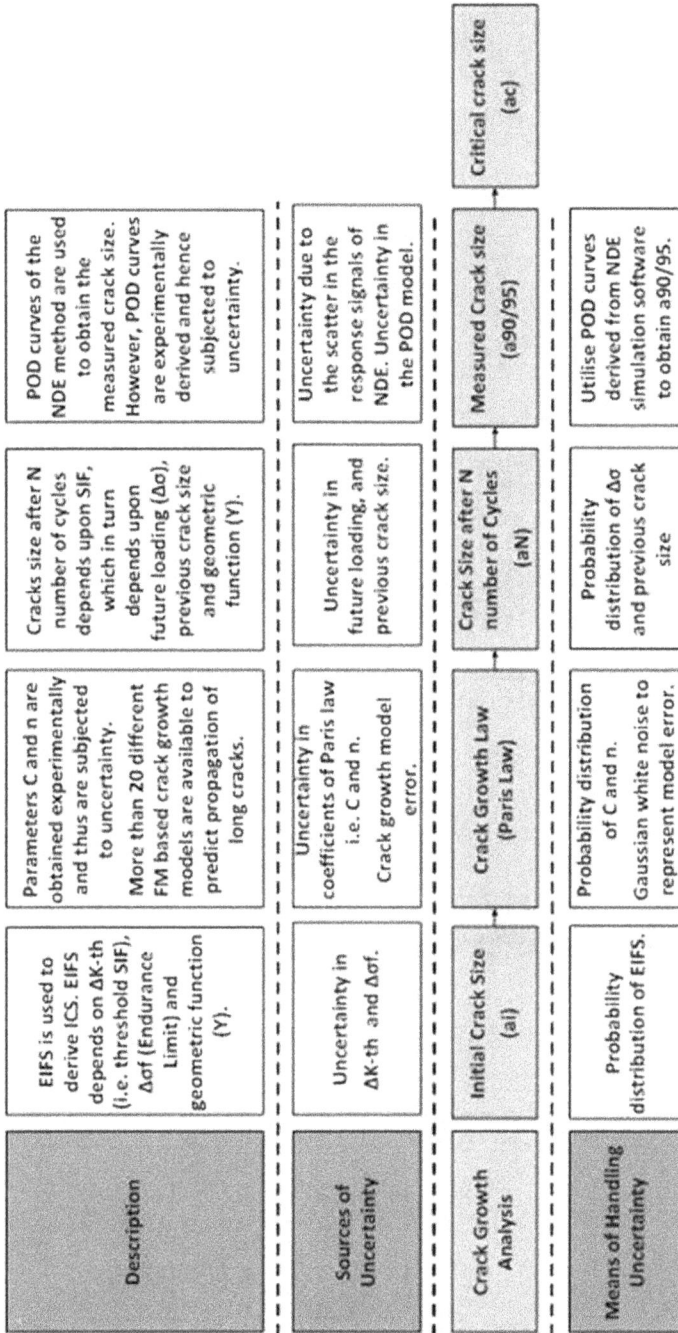

Figure 9.7 Identification of uncertainties during the crack growth analysis of topside piping.

	Initial Crack Size (ai)	Crack Growth Law (Paris Law)	Crack Size after N number of Cycles (aN)	Measured Crack size (a90/95)	Critical crack size (ac)
Description	EIFS is used to derive ICS. EIFS depends on ΔK-th (i.e. threshold SIF), Δσf (Endurance Limit) and geometric function (Y).	Parameters C and n are obtained experimentally and thus are subjected to uncertainty. More than 20 different FM based crack growth models are available to predict propagation of long cracks.	Cracks size after N number of cycles depends upon SIF, which in turn depends upon future loading (Δσ), previous crack size and geometric function (Y).	POD curves of the NDE method are used to obtain the measured crack size. However, POD curves are experimentally derived and hence subjected to uncertainty.	
Sources of Uncertainty	Uncertainty in ΔK-th and Δσf.	Uncertainty in coefficients of Paris law i.e. C and n. Crack growth model error.	Uncertainty in future loading, and previous crack size.	Uncertainty due to the scatter in the response signals of NDE. Uncertainty in the POD model.	
Crack Growth Analysis					
Means of Handling Uncertainty	Probability distribution of EIFS.	Probability distribution of C and n. Gaussian white noise to represent model error.	Probability distribution of Δσ and previous crack size	Utilise POD curves derived from NDE simulation software to obtain a90/95.	

9.4.2 Computational fluid dynamics analysis

Figure 9.8 shows the steps and tools (methodology) used in the present study to estimate the stress. An FSI is performed using ANSYS CFX and ANSYS mechanical. The CAD model depicted in Figure 9.9a has an outer diameter of 219 mm and inner diameter of 168.7 mm for the main pipe. Likewise, the outer diameter for SBP is 60.3 mm and inner diameter is 38.16 mm. Thereafter the CAD model is segmented into fluid region representing the

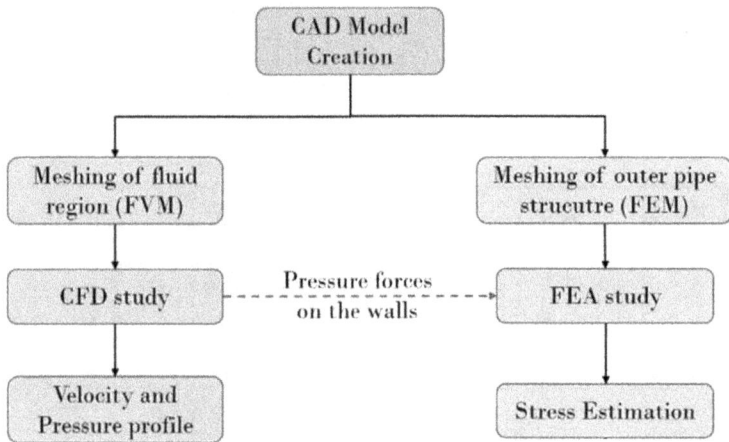

Figure 9.8 Stress prediction methodology.

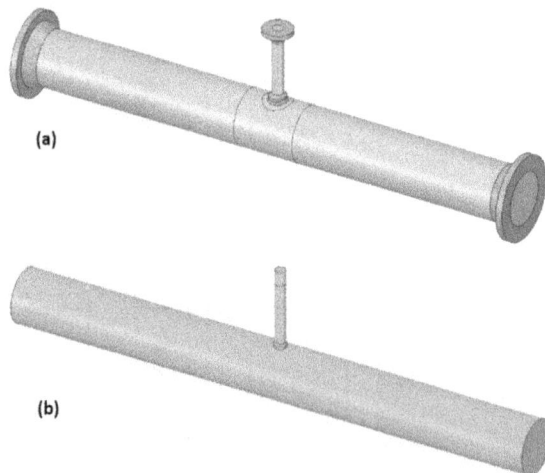

Figure 9.9 CAD model showing (a) pipe and fluid region, (b) complete fluid region.

fluid flow and solid region representing the pipe body, as indicated in Figure 9.9b. The internal volume is meshed using finite volume method (FVM), as shown in Figure 9.10a. While Figure 9.10b depicts the mesh distribution at the confluence of SBP and main piping, likewise the distribution of inflation layers and mesh details are shown in Figure 9.10c. The mesh size was selected after a comprehensive mesh sensitivity study. Twelve layers of inflation with a growth rate of 1.15 are used for CFD meshing.

Finite element method coupled with sensitivity analysis is employed for creating and deciding the mesh size (as shown in Figure 9.11a and Figure 9.11b) for FEA. The fluid (air) incoming velocity is 15 m/s while the downstream pressure (DS) is 30 bar. The parameters of interest in the CFD study are velocity, density, and pressure. The pressure applied by the flowing fluid on the internal walls of the pipe is transferred to the FEA software (as depicted in Figure 9.8) to estimate the stress on the structure arising due to the pressure loads. The maximum internal pressure is 79 MPa, while the external maximum pressure is 48 MPa. It can be seen that the stress is higher on the inner diameter of the pipe compared to the outer diameter. The calculated stress is used as an input while performing RUL assessment.

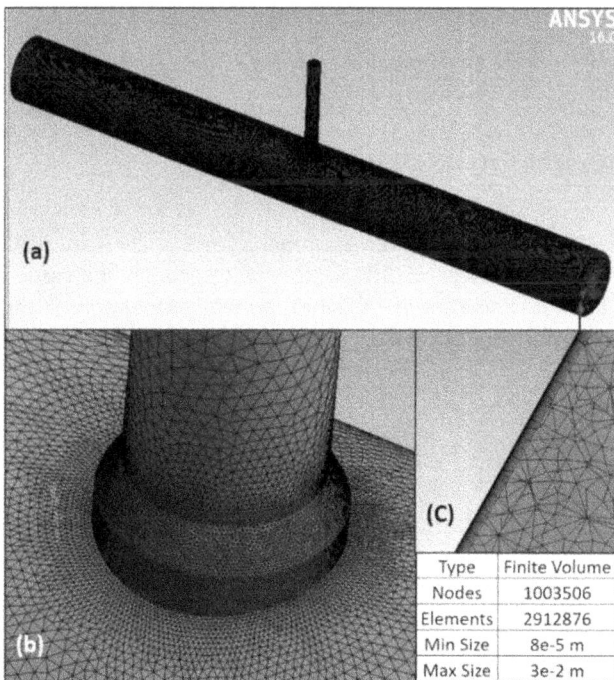

Type	Finite Volume
Nodes	1003506
Elements	2912876
Min Size	8e-5 m
Max Size	3e-2 m

Figure 9.10 Mesh distribution for CFD analysis.

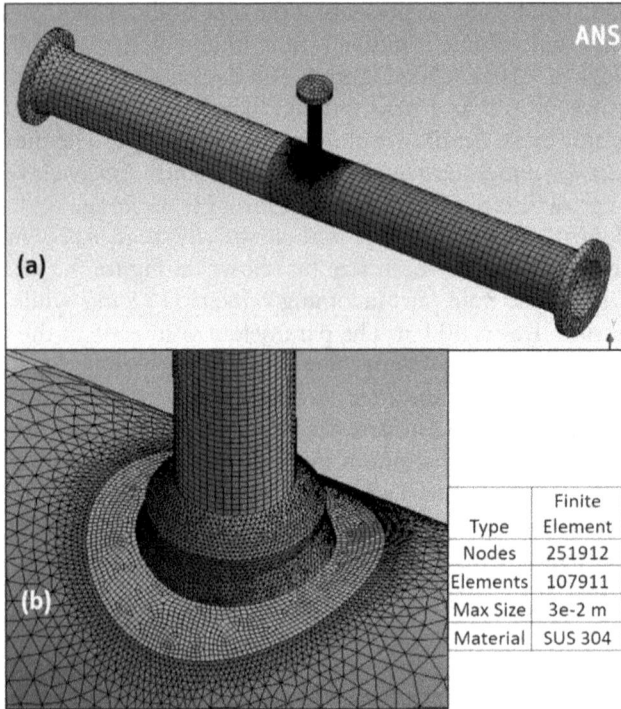

Figure 9.11 Mesh distribution for FE analysis.

9.4.3 Deterministic RUL assessment

During the deterministic RUL prediction, the parameters of the Paris law are fixed, and the values of the parameters are shown in Table 9.1. The minimum detection limit for ultrasonic testing decides the value of ai (Beek et al., 2014), and the value of ac is half of wall thickness for the pipe. The value of geometric function (Y) is taken from the British Standard (BS7901,

Table 9.1 Parameters for deterministic RUL prediction

Parameter Name	Parameter symbol	Parameter value
Initial crack size	ai(mm)	1.5
Material parameter	C	5.21e-13
Material parameter	m	3.5
Remote stress range	$\Delta\sigma$ (MPa)	79
Critical crack size	ac(mm)	11
Geometric function	Y	0.952

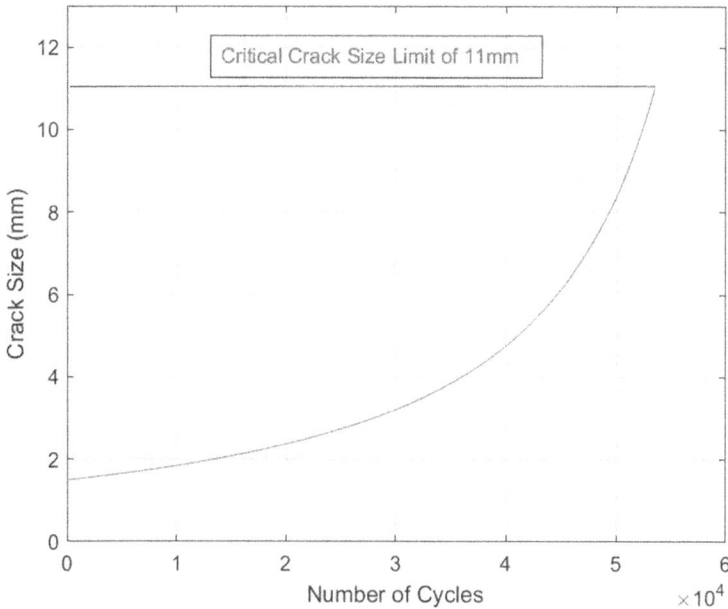

Figure 9.12 Deterministically obtained RUL.

2015) After performing RUL assessment deterministically, the value of RUL estimates to be 5,3549 cycles, as shown in Figure 9.12.

9.4.4 Probabilistic RUL assessment

While performing the RUL estimation probabilistically, the various attributors of uncertainty are taken into consideration, which implies that the different parameters of the Paris law are treated as random variables and are represented by a suitable probability distribution and associated parameters, as shown in Table 9.2. For our analysis, we consider the uncertainty only in three parameters, namely initial crack size, a_i, material parameter C, and $\Delta\sigma$.

Table 9.2 Parameters for probabilistic RUL prediction

Parameter Name	Parameter symbol	Parameter value
Initial crack size	ai(mm)	LogNormal (1.5. 0.05)
Material parameter	C	LogNormal (5.21e-13, 2e-13)
Material parameter	m	3.5
Remote stress range	$\Delta\sigma$ (MPa)	LogNormal (79,5)
Critical crack size	ac(mm)	11
Geometric function	Y	0.952

Furthermore, the modeling error ε_r (whose value in practice is obtained from the experiments) is represented by a Gaussian distribution having mean of zero and standard deviation of 0.05 (Sankararaman et al., 2011).

We have used MCS as means of uncertainty propagation, and 10,000 samples were sampled to obtain the RUL probability distribution, as depicted in Figure 9.13. The mean value of the RUL is equal to 31,813 cycles, while the lower bound 95% confidence interval (CI) was 12,628 and upper bound 95% CI was 66,595 cycles. The associated, CDF, and reliability function are shown in Figure 9.14. As per DNV (DNV-RP-G101, 2010), the first inspection must be performed after one-third of the lower bound of the predicted RUL (i.e. 4,209 cycles), as shown in Figure 9.15. Thus, by using the abovementioned methodology, the unwanted SBPP breakdown due to fatigue can be averted. Furthermore, even more accurate estimation of inspection intervals can be done by employing multi-objective optimization technique, with inspection cost, RUL, and other factors (such as human error) as inputs to the optimization problem. Such a technique will help in striking a balance between structural reliability and inspection cost, eventually leading to augmented process safety and minimization of the total life cycle cost.

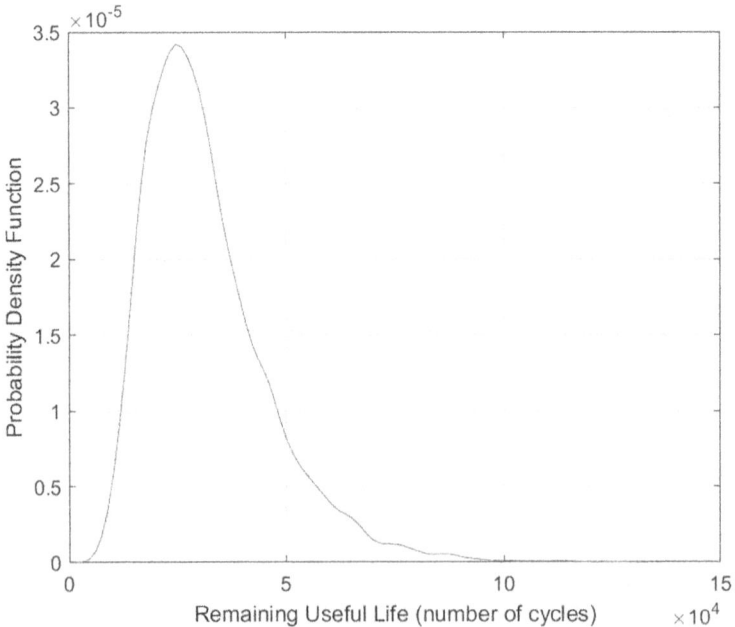

Figure 9.13 Probability density function of RUL of process piping (10,000 samples).

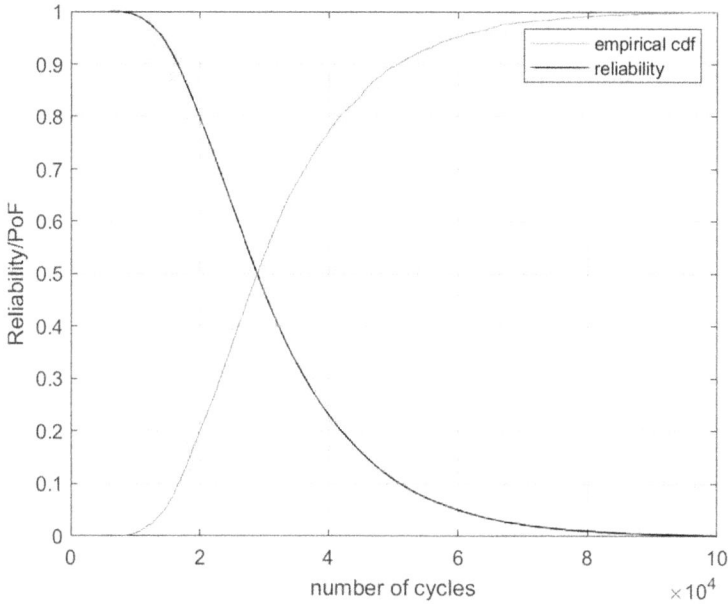

Figure 9.14 Cumulative density function and associated reliability curve of process piping.

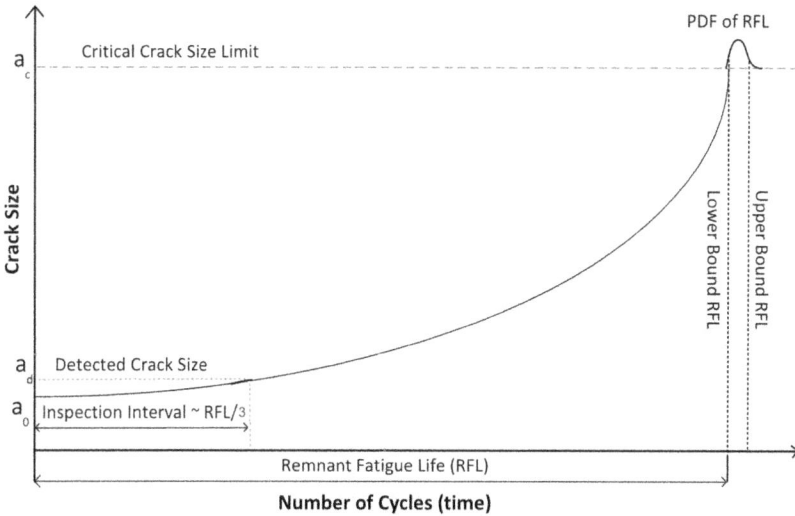

Figure 9.15 Inspection interval derived from RUL.

9.5 CONCLUSION

SBPP failure due to fatigue is common on offshore platforms. In order to avert this failure, it is vital to estimate the RUL of SBPP. In this chapter, authors have laid down the methodology for RUL estimation, where special emphasis is laid on quantification and propagation of the uncertainty in the various parameters of the Paris law. The nominal stress range at the junction of SBPP and MP was obtained using FSI performed in ANSYS. The deterministically estimated RUL was equal to 53,549 cycles, while the same when estimated probabilistically had the mean value of 31,813 cycles with the lower and upper bounds equal to 12,628 and 66,595 cycles, respectively. Finally, inspection interval for SBPP was estimated to be 4,209 cycles by employing method proposed in DNV standard.

REFERENCES

Beek, P.V., Pijpers, R., Macdonald, K.A., Maljaars, J., Lunde, K., Korst, H., and Hansen, F. 2014. A novel high cycle fatigue assessment of small bore side branches: Tailor made acceptable vibration levels based on the remaining life of existing structures. Proceedings of the Pressure Vessels and Piping Conference, Anaheim, California, USA.

BS7910, 2015. *Guide on methods for assessing the acceptability of flaws in metallic structures*. British Standard Institution, London, UK.

Chatterjee, S., and Keprate, A. 2021.Exploratory data analysis of the N-CMAPSS dataset for prognostics. *IEEE International Conference on Industrial Engineering and Engineering Management (IEEM)*, pp. 1114–1121.

DNV, 2020. https://www.dnvgl.com/oilgas/laboratories-testsites/article/tubes-and-piping-are-the-most-failureprone-components.html. Accessed on 09.12.2022.

DNV RP-G101, 2010. *Risk based inspection of offshore topsides static mechanical equipment*. Det Norske Veritas, Høvik, Norway.

EI Guidelines, 2007. *Guidelines for the avoidance of vibration induced fatigue failure in process pipework*. The Energy Institute, London, UK.

Engel, S.J., Gilmartin, B.J., Bongort, K., and Hess, A., 2000. *Prognostics, The real issues involved with predicting life remaining*. IEEE.

Harper, C.B., 2014. Integrity evaluation of small bore connections. 9th Conference of the Energy Frontier Research Center, Vienna, Austria.

HSE RR672, 2008. Offshore hydrocarbon releases 2001–2008. Research Report 672 by Health Safety Executive, Merseyside, UK.

HSE-OTR 002, 2002. *Offshore hydrocarbon releases statistics and analysis*. HID Statistics Report by Health Safety Executive, Merseyside, UK.

Keprate, A. and Ratnayake, R.M.C., 2015. Fatigue and fracture degradation inspection of offshore structures and mechanical items: the state of the art. Proceedings of the International Conference on Offshore Mechanics and Arctic Engineering (OMAE 2015), St. John's, Newfoundland, Canada.

Keprate, A. and Ratnayake, R.M.C., 2016. Selecting a modeling approach for predicting remnant fatigue life of offshore topside piping. *IEEE International*

Conference on Industrial Engineering and Engineering Management (IEEM), Bali, Indonesia, pp. 1407–1411.

Keprate, A., Ratnayake, R.M.C., and Sankararaman, S., 2017. Minimizing hydrocarbon release from offshore piping by performing probabilistic fatigue life assessment. *Process Safety and Environmental Protection*, 106, 34–51.

MATLAB®, 2022. https://se.mathworks.com/help/predmaint/ug/rul-estimation-using-rul-estimator-models.html. Accessed on 09.12.2022.

Sankararaman, S., Ling, Y., Shantz, C. and Mahadevan, S., 2011. Uncertainty quantification and model validation of fatigue crack growth prediction. *Engineering Fracture Mechanics*, 78(7), 1487–1504.

Swindell, R., 2003. *Vibration fatigue in process pipework: A risk based assessment methodology*. Bureau Veritas, Southampton, UK.

Chapter 10

Minimization of joint angle jerk for industrial manipulator based on prognostic behaviour

J. Vaishnavi and Bharat Singh
Malaviya National Institute of Technology, Jaipur, India

Ankit Vijayvargiya
Swami Keshvanand Institute of Technology, Management & Gramothan, Jaipur, India

Rajesh Kumar
Malaviya National Institute of Technology, Jaipur, India

CONTENTS

10.1 INTRODUCTION

Robots are designed to do monotonous activities autonomously. Several types of robots are developed based on their usage (Singh, Kumar and Singh 2022). They have transformed many industries including manufacturing, transport (Qi, Wang and Shan 2018), space explorations (Gao and Chien 2017), intricate surgeries (Kahrs, Rucker and Choset 2015; Jayaswal, Palwalia and Kumar 2021), military (Payal et al. 2021; Reis et al. 2021), etc.

In recent times, prognostics and health management (PHM) for engineering systems, such as machine learning-enabled system, failure detection,

DOI: 10.1201/9781003434849-10

sensing techniques, and real-time health monitoring of the complex engineering system under the real-time operating conditions, has become an essential component (Kim, An and Choi 2017). Robots have mechanical moving parts which are prone to damage or failure. It is vital to ensure that the robotic systems operate in such a way that they have a longer operating life. During the COVID-19 pandemic, robotic applications made a significant contribution (Wang and Wang 2021), especially robot-assisted surgeries, due to their critical role in removing human error, lowering complication rates, and increasing accuracy, their immunity to infection, and their efficient disinfection routine. These surgical robots include manipulator arms that track the appropriate operating trajectories and aid in procedures (Jayaswal, Palwalia and Kumar 2021). Applications with robots-human interaction or robots working in hazardous environment require proper control techniques with minimum vibrations to prevent any adverse condition. For example, the control of surgical robots has to ensure that the operation of robot is going smoothly to avoid any danger to patient. Also, minimizing repeated abrupt motion of links of robots will increase operating life of motors in robots' joints.

A variety of control techniques, namely torque and kinematic control, are designed to conduct operations on the manipulators with high accuracy, precision, and stability. Due to the mathematical model inaccuracy, control mechanisms face accuracy issues, resulting in a dangerous situation.

Several researchers have devised the inverse kinematics (IK) approach to overcome these issues. The joint angles of an n-joint robot are calculated in the IK problem so that the robot end-effector (EE) achieves the desired position with a specific orientation. The difficulty of solving the IK problem is determined by the robot arm's geometry and the non-linearity of the joint to Cartesian space mapping. The accuracy of the IK solution is critical for the successful operation of the robot duties. Multiple solutions, singularity, physical constraint violation, and redundancy are all issues that arise when addressing the IK problem. Conformal geometric algebra (geometric approach) (Wang et al. 2018), Grobner Bases (algebraic technique) (Horigome, Terui and Mikawa 2020), and combined optimization (numerical approach) (Wang and Chen 1991) are insufficient and slow as per literature. The intelligent techniques can be employed to find the solution (El-Sherbiny, Elhosseini and Haikal 2018a), even when there is a possibility of the existence of multiple solutions. Nowadays, intelligent techniques like meta-heuristic algorithms are highly preferred to solve complex problems. These algorithms optimize the objective functions to give the best fitness values. In literature, researchers have optimized objective function involving error between the actual EE position and the desired EE position using analytical, convex optimization techniques and compact differential evolution algorithm with cyclic coordinate descent (CCD) (Kircanski and Petrovic 1991; Marić et al. 2020; Yotchon and Jewajinda 2021). Researchers have also optimized the EE position error function using meta-heuristic techniques (Çavdar, Mohammad and Milani 2013;

Dereli and Köker 2018, 2020a, 2020b; Hernandez-Barragan et al. 2018; Kumar, Murali and Srikanth 2018; Wu, Shi and Wang 2019) such as modified artificial bee colony (ABC), inertia weight particle swarm optimization (PSO) variants, firefly algorithm (FFA), invasive weed optimization (IWO), quantum PSO (QPSO), FFA, genetic algorithm (GA), hybrid GA-Nelder mead technique, extreme learning machine (ELM) based on PSO, ABC, FFA, IWO, PSO, for various degrees of freedom (DoF) robotic manipulators. EE position and orientation plays an important role in trajectory tracking due to which several research works were done on IK solution of different DoFs robotic manipulators taking both EE position and orientation error into consideration. IK of manipulators with 3 to 7 DoFs was solved using functional joint control, numerical solution (Poon and Lawrence 1988; Wang and Chen 1991), and meta-heuristic algorithms such as PSO, ANFIS, ANN, GA, ABC and its variants, and PSO and its variants such as sine-cosine PSO, fully resampled PSO (FRPSO), DPSO, differential evolution (DE), PSO, GA, and mutating PSO (Alkayyali and Tutunji 2019; Collinsm and Shen 2017; El-Sherbiny, Elhosseini and Haikal 2018a, 2018b; Jin and Zhai 2020; Junior et al. 2018; Khan, Abbasi and Lee 2020; Li et al. 2021; Reyes and Gardini 2019; Tabandeh, Clark and Melek 2006; Umar et al. 2019). Furthermore, to reduce the joint angle displacement in a trajectory, researchers have implemented meta-heuristic techniques such as QPSO, PSO, ABC, grey wolf optimization (GWO), modified GA, DE, PSO, and modified version of DE/PSO to minimize an objective function involving joint angle displacement term along with either EE position error or orientation error, and solved the IK problem for 4-, 5-, 7-, and 10-DoF robotic manipulators (Cao et al. 2021; Huang, Chen and Wang 2012; Kumar et al. 2019; Nearchou 1998; Nguyen, Nguyen and Nguyen 2020). However, most of these researchers have analysed these algorithms over the complete search space of the joint angle limits. Due to this, the IK solution obtained for consecutive points in the trajectory may result in discontinuous angles. These solutions could be highly varying in magnitude, resulting in sudden jerky motion of the links throughout the trajectory. In this chapter, the IK of a trajectory of a 5-DoF robotic manipulator is computed using the PSO algorithm with two different objective functions. The first objective function is optimized using PSO with the entire search space for initialization, and then the second objective function is optimized with a reduced search space. The performance of PSO is evaluated based on the errors and continuity/discontinuity of joint angles obtained from the IK solutions with both the optimization functions. Major contributions of this work are as follows:

- Develop the objective function which integrates the EE position and orientation with the joint angles.
- Obtain continuous and smooth joint angles by modifying the search space of optimization problem close to previous solution for PSO algorithm.

This chapter is organized as follows: Section 10.2 describes a 5-DoF robot manipulator model and mathematical expression for forward kinematic using Denavit–Hartenberg (DH). Section 10.3 describes two different objective functions which are minimized by PSO. Section 10.4 illustrates the results of this study. Finally, Section 10.5 provides the conclusion.

10.2 SYSTEM DESCRIPTION

A robotic manipulator consists of links, joints, and an EE. The number of independent joint variables necessary to identify the position of all the links of the robot in space is known as the number of DoFs. The robotic manipulator selected for this research, as shown in Figure 10.1, has five degrees of freedom. The joints in this manipulator are revolute joints.

Table 10.1 shows the values of the manipulator's geometrical parameters such as length of the links and the joint angle limits. Geometric modelling of robots is essential for the control of EE with respect to manipulator's base. The DH parameters that give the relationship between the EE coordinate frame and base frame of the robot structure are defined below.

Joint offset, d_i: offset along z_{i+1} to the common normal
Joint angle, θ_i: angle about z_i, from x_i to x_{i+1}
Link length, a_i: length of the common normal
Twist angle, α_i: angle about common normal x_{i+1}, from z_i axis to z_{i+1} axis

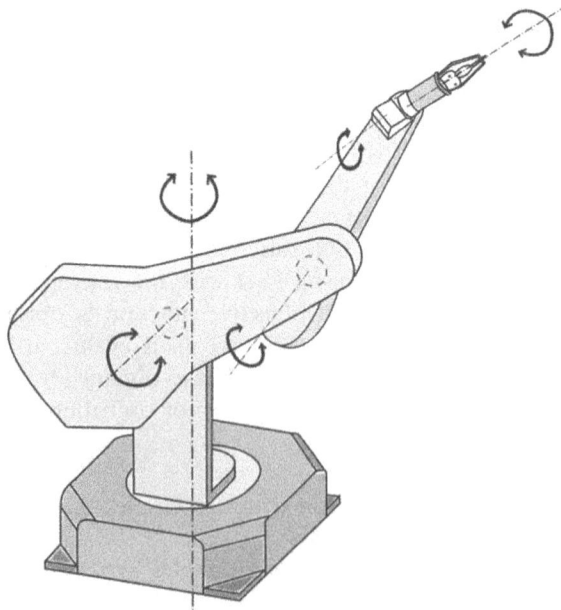

Figure 10.1 5 DoF robotic manipulator.

Table 10.1 Parameter values for 5-DoF robotic arm

Symbol	Parameter	Value	Unit
l_1	Base link length	38	cm
l_2	First link length	28	cm
l_3	Second link length	28	cm
l_4	Third link length	8	cm
ϕ_1	Ist joint angle	0 to 2π	rad
ϕ_2	2nd joint angle	0 to $\pi/2$	rad
ϕ_3	3rd joint angle	$-\pi/2$ to 0	rad
ϕ_4	4th joint angle	$-\pi/2$ to $\pi/2$	rad
ϕ_5	5th joint angle	0 to π	rad

Table 10.2 DH Parameters of 5-DoF robotic arm

Link	a_{i-1}	α_i	d_i	θ_i
1	0	0	l_1	ϕ_1
2	l_2	$-\pi/2$	0	ϕ_2
3	l_3	0	0	ϕ_3
4	l_4	$-\pi/2$	0	ϕ_4
5	0	$\pi/2$	0	ϕ_5

The DH parameters for the robot in Figure 10.1 are shown in Table 10.2. Forward kinematics gives the expression for EE position and orientation in terms of joint angles. For the selected manipulator, the forward kinematic equations are given by Eq. (10.1):

$$\left.\begin{aligned}
r &= l_2 \cos\phi_2 + l_3 \cos(\phi_2 + \phi_3) + l_4 \cos(\phi_2 + \phi_3 + \phi_4) \\
X_e &= r \cos\phi_1 \\
Y_e &= r \sin\phi_1 \\
Z_e &= l_1 + l_2 \sin\phi_2 + l_3 \sin(\phi_2 + \phi_3) + l_4 \sin(\phi_2 + \phi_3 + \phi_4) \\
\phi_{X_e} &= \phi_5 \\
\phi_{Y_e} &= \phi_2 + \phi_3 + \phi_4 \\
\phi_{Z_e} &= \tan^{-1}(Y_e / X_e) = \phi_1
\end{aligned}\right\} \quad (10.1)$$

10.3 ALGORITHMS AND OBJECTIVE FUNCTIONS

This section discusses the objective functions which include the error minimization of position and orientation of EE. The reasoning behind the modified solution search space is briefly discussed. Apart from this, a meta-heuristic optimization algorithm, i.e., PSO, is also discussed.

10.3.1 Objective function

Initially, for the purpose of solving IK, an objective function is selected such that it includes the error between the desired EE position and orientation, which is given by Eq. (10.2):

$$F_1 = \sqrt{\Delta x^2 + \Delta y^2 + \Delta z^2 + \Delta \phi_{X_e}^2 + \Delta \phi_{Y_e}^2 + \Delta \phi_{Z_e}^2} \tag{10.2}$$

Here, $\Delta x = X_e - x_{ik}$, $\Delta y = Y_e - y_{ik}$, $\Delta z = Z_e - z_{ik}$, $\Delta \phi_{Xe} = \phi_{Xe} - \phi_{xik}$, $\Delta \phi_{Ye} = \phi_{Ye} - \phi_{yik}$, $\Delta \phi_{Ze} = \phi_{Ze} - \phi_{zik}$.

X_e, Y_e, Z_e, ϕ_{Xe}, ϕ_{Ye}, and ϕ_{Ze} are the desired EE position and orientation, and x_{ik}, y_{ik}, z_{ik}, ϕ_{xik}, ϕ_{yik}, and ϕ_{zik} are the position and orientation after substituting IK solution obtained from optimization techniques.

10.3.2 Modified objective function

After obtaining the solution of IK using PSO to minimize the objective function F1 in the complete search space of joint angles within the constraints, a new objective function is defined in such a way that it contains EE position and orientation error along with the difference between previous IK solution and the current solution. The modified objective function is represented by Eq. (10.3):

$$F_2 = \sqrt{\Delta x^2 + \Delta y^2 + \Delta z^2} + \sqrt{\Delta \phi_{X_e}^2 + \Delta \phi_{Y_e}^2 + \Delta \phi_{Z_e}^2} + \sqrt{\sum_{i=1}^{N}\left(\phi_{i,p} - \phi^*_{i,p}\right)^2} \tag{10.3}$$

Here, Δx, Δy, Δz, $\Delta \phi_{Xe}$, $\Delta \phi_{Ye}$, and $\Delta \phi_{Ze}$ are as defined in Subsection 10.3.1. $\phi_{i,p}$ is the joint angle i at the p-th trajectory point, $\phi^*_{i,p}$ is the best solution (joint angle i) obtained by PSO for the previous (p-1) trajectory point, and N is the dimension of the solution. Here, $N = 5$ for the joint angles of manipulator.

10.3.3 Particle swarm optimization

James Kennedy and Russell Eberhart proposed PSO algorithm in 1995, taking inspiration from the behaviour of a flock of birds and fish schooling (Kennedy and Eberhart 1995). An optimization problem is solved by a swarm of particles having initial position and velocity. After that, in every iteration they move with certain velocity and update their position till the termination criteria is met. The termination criteria can be number of iterations, obtaining optimal solution or continuously repeating solution. The optimal solution is the one that produces the best objective function value (fitness) in the last iteration. The velocity v_i and position x_i of each particle i is updated in each iteration $t + 1$ using Eqs. (10.4) and (10.5), respectively,

$$v_i^{t+1} = w.v_i^t + c_1.rand_2().\left(P_{best,i} - x_i^t\right) + c_2.rand_2().\left(G_{best} - x_i^t\right) \qquad (10.4)$$

$$x_i^{t+1} = x_i^t + v_i^{t+1} \qquad (10.5)$$

In the above equations, $P_{best,i}$ is the position of the i-th particle's best solution corresponding to its best fitness value, and G_{best} is the global best position of particle with the best fitness value in the entire swarm population. c_1 and c_2 are the accelerating coefficients, called cognitive learning rate and social learning rate, respectively, which help decide the influence of $P_{best,i}$ and G_{best} optimal positions on the particles velocity. Low values of c_1 and c_2 are taken to obtain more exploration and smooth particle trajectories. High values result in sudden particle motion due to high acceleration. w is the inertia weight that is responsible for the control of particle's momentum during their movement in the search space. Lower value of w results in quick change in particle's direction of motion (Kumar and Fozdar 2017). For the objective function F_1, the search space that is considered is the entire range of joint angles, as given in Table 10.1. For the modified objective function F_2, the search space of the optimization problem, i.e., the joint angles' constraints, has been modified, as shown in Figure 10.2. The initialization of

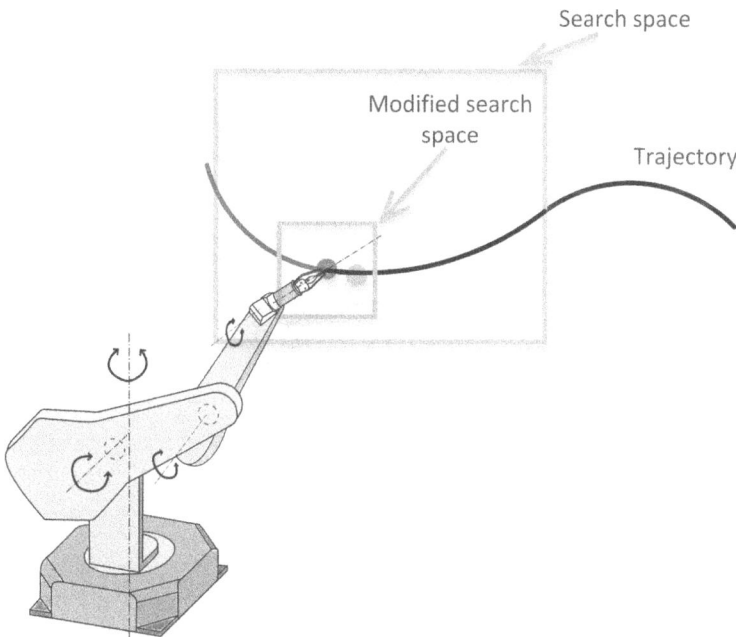

Figure 10.2 Modification in search space.

the PSO particles in successive trajectory points is around the previous trajectory point resulting in highly reduced search space. In case the modified search space of any joint angle violates the joint angle limits for a point in trajectory, the search space is reset to original limits, and initialization of particles is done accordingly. These steps ensure that the particles in PSO search around the previous joint angles, resulting in continuous IK solution angles (with very low variation) for the consecutive points of trajectory, and at the same time satisfy the constraints.

10.4 RESULTS AND DISCUSSION

The desired trajectory is defined using the EE's position and orientation. In this research work, PSO solves the IK problem by minimizing the two objective functions. The termination criteria for PSO considered in this work is the setting for the total number of iterations, i.e., 400.

10.4.1 Performance evaluation for objective function F_1

At first, PSO optimizes the objective function F_1, which is given by Eq. (10.2). Figure 10.3a shows the reference trajectory and the actual position of EE (X-Y-Z coordinates) after the objective function has been minimized by PSO. Figures 10.3b–d show the reference EE orientation and actual orientation after substituting IK solution in (10.1). As seen from the Figure 10.3, the IK solution of the trajectory doesn't follow the desired trajectory and has

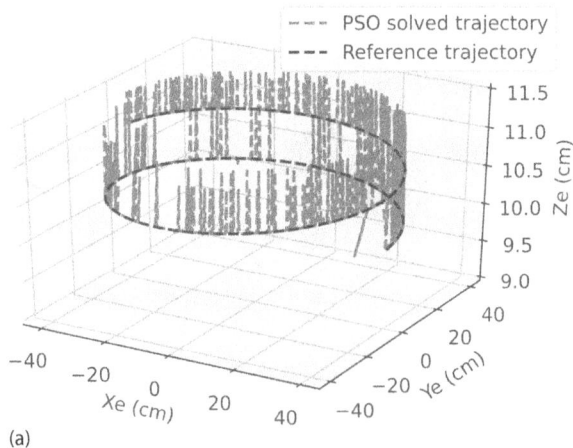

(a)

Figure 10.3 EE position and orientation with objective function F_1 and complete search space. (a) X_e, Y_e, Z_e coordinates.

(b)

(c)

(d)

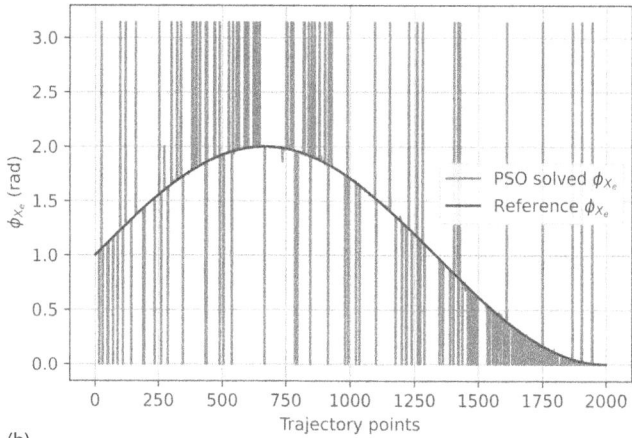

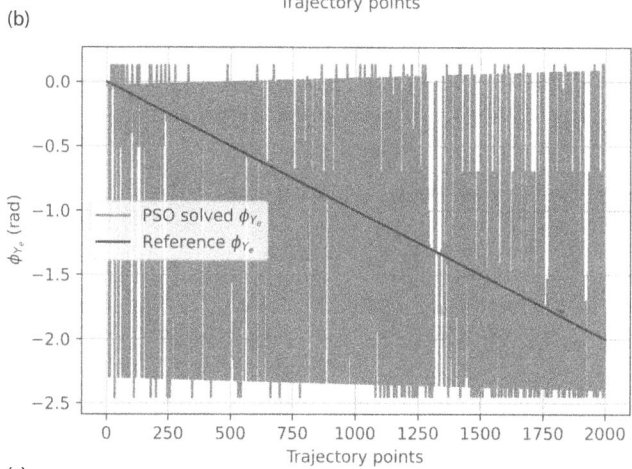

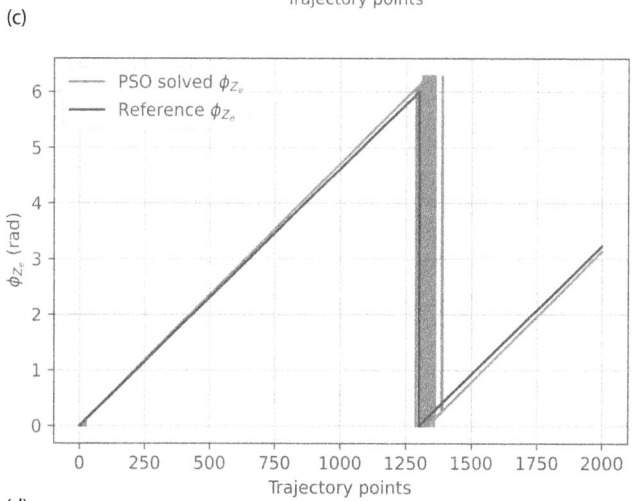

Figure 10.3 (Continued) (b) ϕ_{Xe}. (c) ϕ_{Ye}. (d) ϕ_{Ze}.

inaccuracy in tracking. Since the entire search space has been used for the implementation of the PSO algorithm to solve the IK of the manipulator, the successive points of the trajectory don't necessarily have continuous joint angle solution. As seen from Figure 10.5, the joint angles ϕ_1, ϕ_2, ϕ_3, ϕ_4, and ϕ_5 are discontinuous; therefore, they cause vibrations during motion and make the manipulator infeasible for practical applications. Total error is calculated using Eq. (10.6), where PE and OE are the position and orientation error respectively, p denotes the point in the trajectory, and T is the total number of points in the trajectory. This total error from Eq. (10.6) computed for all the points in the trajectory is 0.030000877.

$$E = \frac{1}{T}\sqrt{\sum_{p=1}^{T}\left(PE_p^2 + OE_p^2\right)} \tag{10.6}$$

The position and orientation error in the total error is found out to be 0.015924407 cm and 0.025425694 radians, respectively.

10.4.2 Performance evaluation for modified objective function

Since the trajectory tracking has accuracy issues and jerky motion, and joint angles continuity is not maintained with previous optimization problem, prognostics might detect future failure possibilities on the account of joint breakage or component damage. Therefore, as an opportunity to take preventive action, the trajectory tracking is repeated with some variations. PSO, on the other hand, now optimizes the modified objective function F_2, given by Eq. (10.3). For the first trajectory point, the entire search space is utilized. For the next consecutive points, the search space for the initialization of the PSO particles is reduced around the previous joint angle values ($\phi_{i-1} - 0.03722 \leq \phi_i \leq \phi_{i-1} + 0.03722$) radians. Figure 10.4 shows the reference trajectory and the actual position of EE (X-Y-Z coordinates) after PSO minimizes the objective function. Figures 10.4b to 10.4d show the reference EE orientation and actual orientation after PSO solves the IK problem. The EE in this approach follows the desired trajectory more accurately than that in the first approach with function F_1. Here, the particles of PSO algorithm search for optimal solution in a modified search space, because of which the joint angles ϕ_1, ϕ_2, ϕ_3, ϕ_4, and ϕ_5 are almost continuous, as seen in Figure 10.6. The overall reduced joint angle displacement in the complete trajectory tracking suggests that the objective function and search space modification has helped the manipulator perform better concerning the links' vibrations during motion. Total error in this setting is calculated using Eq. (10.6) and is equal to 0.023468741, out of which the position and orientation error amounts to 0.003722472 cm and 0.023171642 radians, respectively.

(a)

(b)

(c)

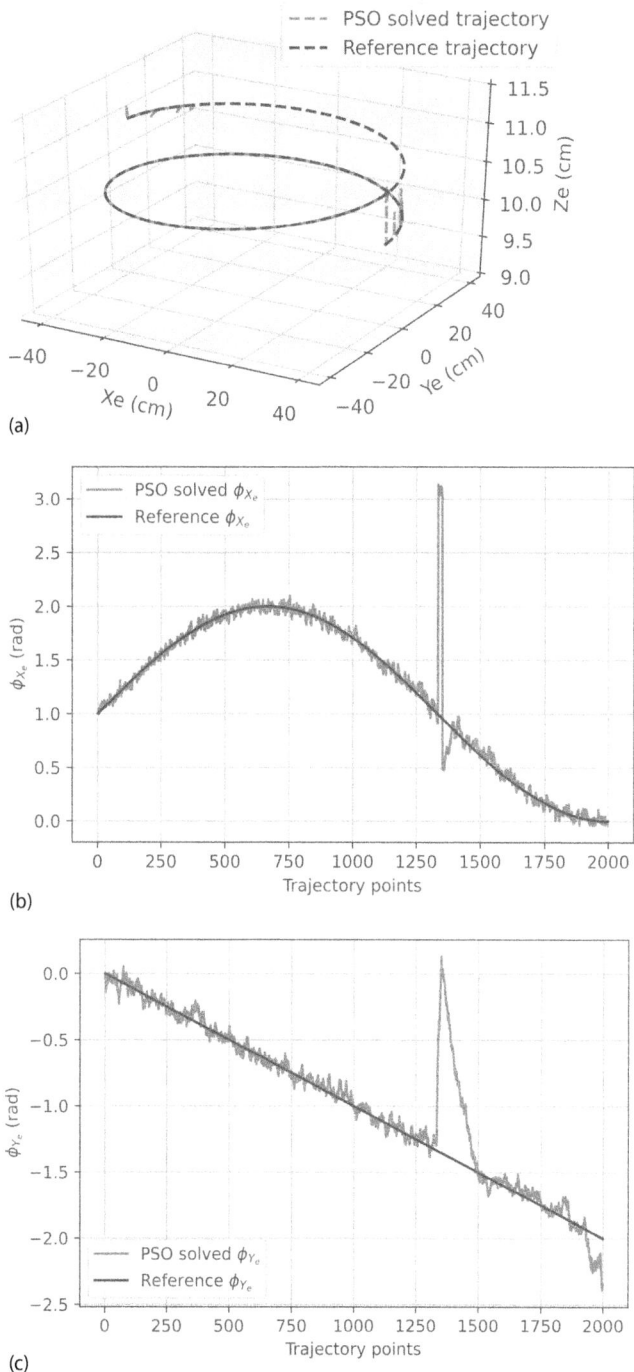

Figure 10.4 EE position and orientation with objective function F_2 and reduced search space. (a) X_e, Y_e, Z_e coordinates. (b) ϕ_{Xe}. (c) ϕ_{Ye}. (d) ϕ_{Ze}.

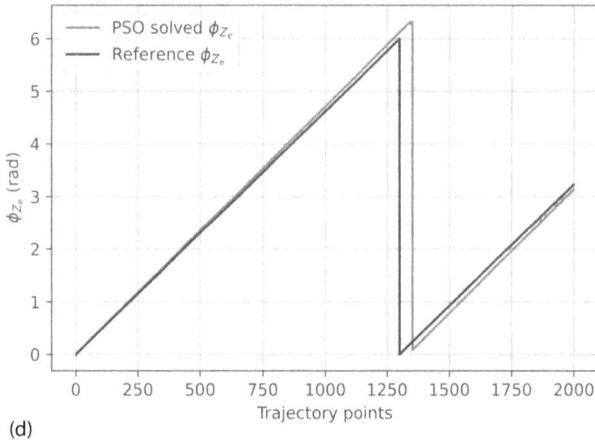

(d)

Figure 10.4 (Continued) (d) ϕ_{Ze}.

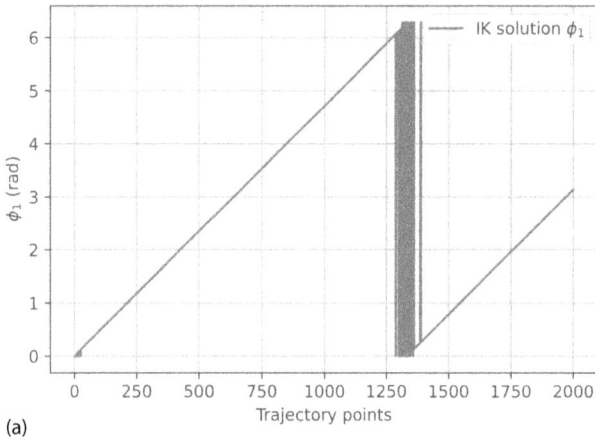

(a)

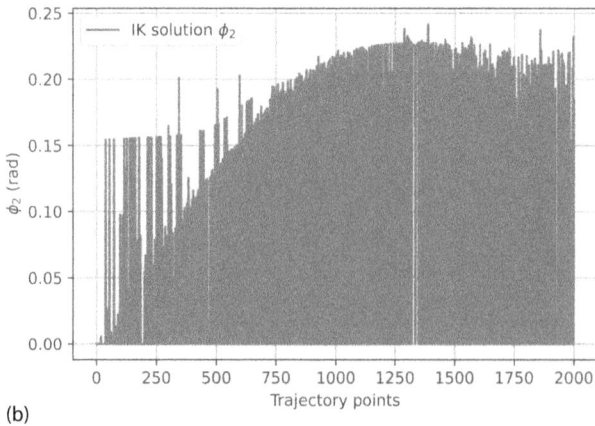

(b)

Figure 10.5 Joint angles when objective function F_1 and complete search space are used. (a) ϕ_1. (b) ϕ_2.

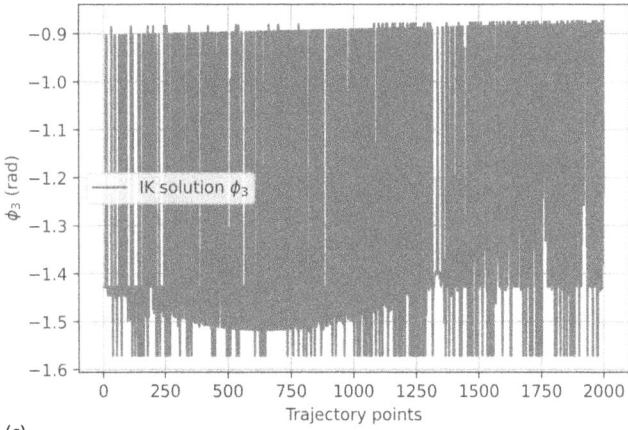

(c)

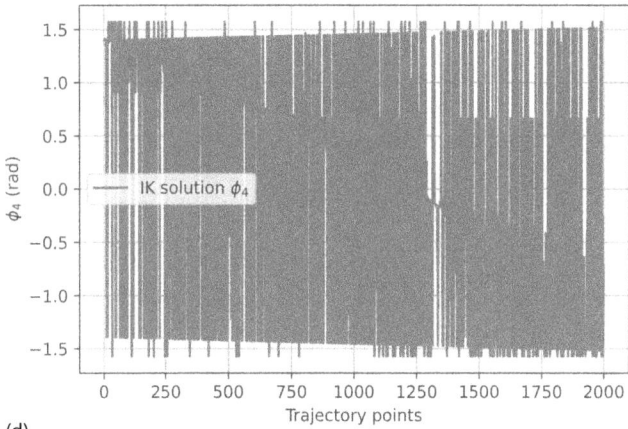

(d)

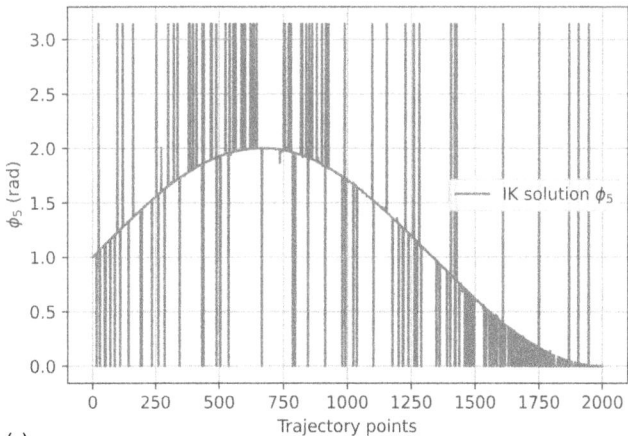

(e)

Figure 10.5 (Continued) (c) ϕ_3. (d) ϕ_4. (e) ϕ_5.

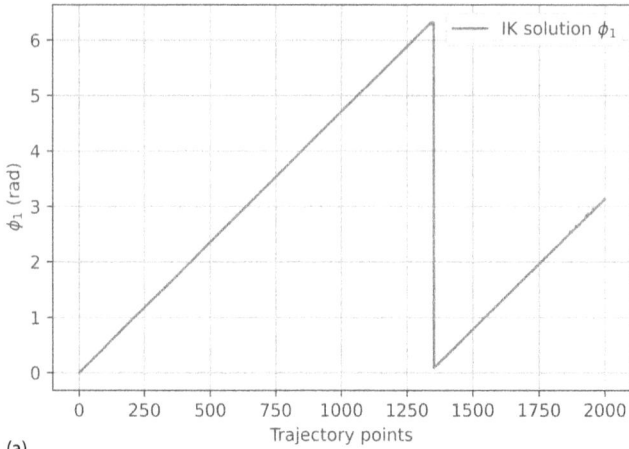

(a)

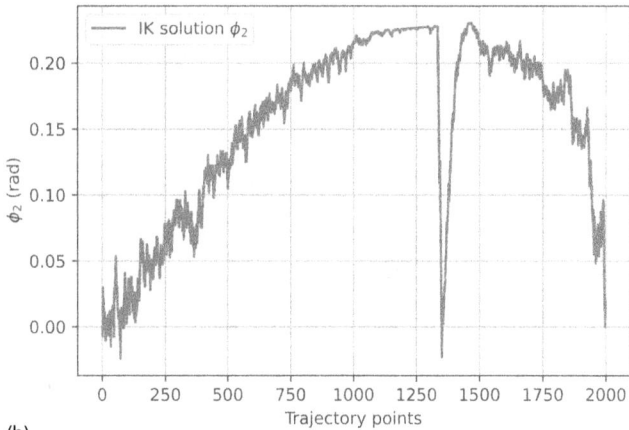

(b)

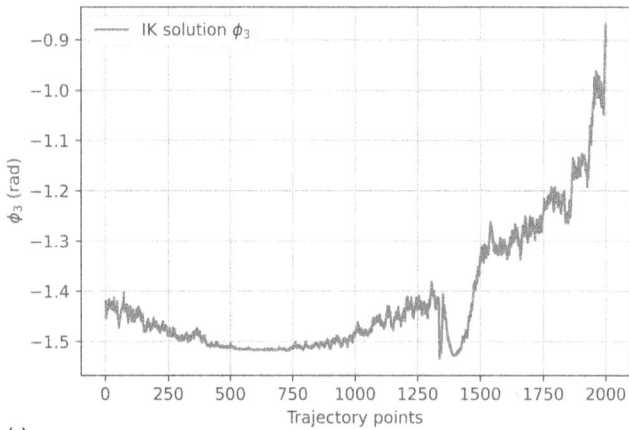

(c)

Figure 10.6 Joint angles when objective function F_2 and reduced search space are used. (a) ϕ_1. (b) ϕ_2. (c) ϕ_3.

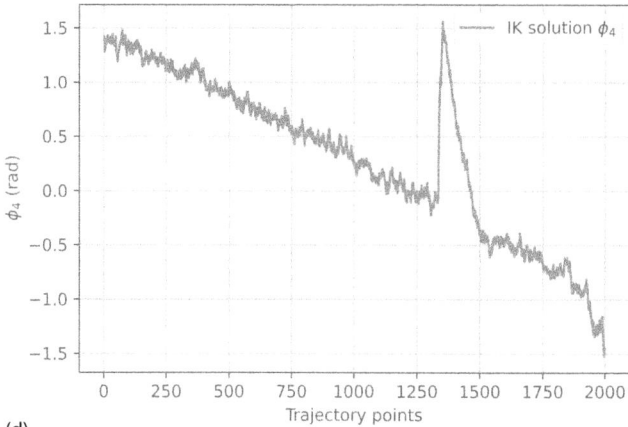

(d)

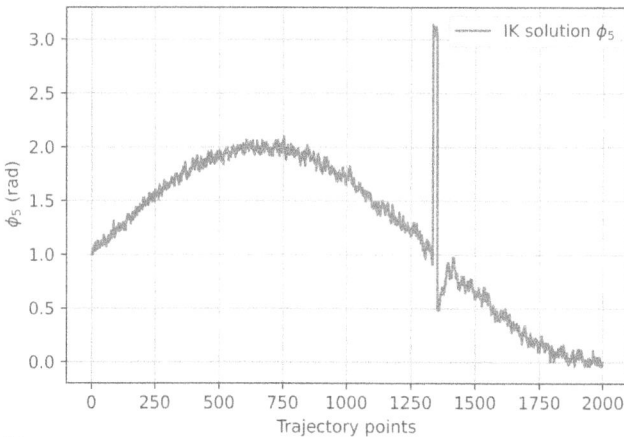

(e)

Figure 10.6 (Continued) (d) ϕ_4. (e) ϕ_5.

10.5 CONCLUSION

The robotic applications are becoming increasingly popular nowadays due to their advantages. In numerous applications, robots are required to move smoothly without any continuous vibrations in the links. The goal of this study is to improve the trajectory tracking of a 5-DoF robotic manipulator by minimizing the sudden jerky motion of the links. This is analysed by optimizing two objective functions by PSO, and simultaneously reducing the search space around the trajectory points. It is observed that the modified objective function and reduced search space for PSO results in accurate tracking. Moreover, the joint angles are continuous and smoother as compared to the results obtained from the first objective function. Performance recorded with the second objective function is better than the first one in terms of total position and orientation error, i.e., the second and first approaches have 0.023468741 and 0.030000877, respectively. The control of robotic system based on the prognostics behaviour ensures proper, safe, steady, and smooth motion of manipulator, along with extended operating life of components in robots. More algorithms can be implemented for further study and trajectory control to obtain smooth motion of manipulator.

REFERENCES

Alkayyali, Malek, and Tarek A. Tutunji. "PSO-based Algorithm for Inverse Kinematics Solution of Robotic Arm Manipulators." *In 2019 20th International Conference on Research and Education in Mechatronics (REM)*. Wels: IEEE, 2019. 1–6.

Cao, Yuting Wenjie Wang, Liping Ma, and Xiaohua Wang. "Inverse kinematics solution of redundant degree of freedom robot based on improved quantum particle swarm optimization." *In 2021 IEEE 7th International Conference on Control Science and Systems Engineering (ICCSSE)*. Qingdao, China: IEEE, 2021. pp. 68–72.

Çavdar, Tuğrul, M. Mohammad, and R. Alavi Milani. "A new heuristic approach for inverse kinematics of Robot Arms." *Advanced Science Letters* 19, no. 1 (2013): 329–333.

Collinsm, Thomas Joseph, and Wei-Min Shen. "Particle swarm optimization for high-DOF inverse kinematics." *In 2017 3rd international conference on control, automation and robotics (ICCAR)*. Nagoya, Japa: IEEE, 2017. pp. 1–6.

Dereli, Serkan, and Raşit Köker. "IW-PSO approach to the inverse kinematics problem solution of a 7-DOF serial robot manipulator." *Sigma Journal of Engineering and Natural Sciences* 36, no. 1 (2018): 77–85.

Dereli, Serkan, and Raşit Köker. "Calculation of the inverse kinematics solution of the 7-DOF redundant robot manipulator by the firefly algorithm and statistical analysis of the results in terms of speed and accuracy." *Inverse Problems in Science and Engineering* 28, no. 5 (2020a): 601–613.

Dereli, Serkan, and Raşit Köker. "A meta-heuristic proposal for inverse kinematics solution of 7-DOF serial robotic manipulator: Quantum behaved particle swarm algorithm." *Artificial Intelligence Review* 53, no. 2 (2020b): 949–964.

El-Sherbiny, Ahmed, Mostafa A. Elhosseini, and Amira Y. Haikal. "A comparative study of soft computing methods to solve inverse kinematics problem." *Ain Shams Engineering Journal* 9, no. 4 (2018a): 2535–2548.

El-Sherbiny, Ahmed, Mostafa A. Elhosseini, and Amira Y. Haikal. "A new ABC variant for solving inverse kinematics problem in 5 DOF robot arm." *Applied Soft Computing* 73 (2018b): 24–38.

Gao, Yang, and Steve Chien. "Review on space robotics: Toward top-level science through space exploration." *Space Robotics* 2, no. 7 (2017).

Hernandez-Barragan, Jesus, Carlos Lopez-Franco, Cecilia Antonio-Gopar, Alma Y. Alanis, and Nancy Arana-Daniel. "The inverse kinematics solutions for robot manipulators based on firefly algorithm." *In 2018 IEEE Latin American Conference on Computational Intelligence (LA-CCI)*. Gudalajara, Mexico: IEEE, 2018. pp. 1–5.

Horigome, Noriyuki, Akira Terui, and Masahiko Mikawa. "A design and an implementation of an inverse kinematics computation in robotics using gröbner bases." *International Congress on Mathematical Software*. Cham: Springer, 2020, pp. 3–13.

Huang, Hsu-Chih, Chien-Po Chen, and Pei-Ru Wang. "Particle swarm optimization for solving the inverse kinematics of 7-DOF robotic manipulators." *In 2012 IEEE international conference on systems, man, and cybernetics (SMC)*. Seoul, Korea (South): IEEE, 2012. pp. 3105–3110.

Jayaswal, Kuldeep, D. K. Palwalia, and Sandeep Kumar. "Performance investigation of PID controller in trajectory control of two-link robotic manipulator in medical robots." *Journal of Interdisciplinary Mathematics* 24, no. 2 (2021): 467–478.

Jin, Fei, and Junyong Zhai. "SCAPSO-based inverse kinematics method and its application to industrial robotic manipulator." *In 2020 39th Chinese Control Conference (CCC)*. Shenyang, China: IEEE, 2020. pp. 5933–5938.

Junior, Jefferson de Lima Silveira, Raphael Cardoso de Oliveira Jesus, Lucas Molina, Elyson Adan Nunes Carvalho, and Eduardo Oliveira Freire. "FRPSO: Inverse kinematics using fully resampled particle swarm optimization." *In 2018 Latin American Robotic Symposium, 2018 Brazilian Symposium on Robotics (SBR) and 2018 Workshop on Robotics in Education (WRE)*. João Pessoa, Brazil: IEEE, 2018. pp. 402–407.

Kahrs, Jessica Burgner, D. Caleb Rucker, and Howie Choset. "Continuum robots for medical applications: A survey." *In IEEE Transactions on Robotics* 31, no. 6 (2015): 1261–1280.

Kennedy, James, and Russell Eberhart. "Particle swarm optimization." *In Proceedings of ICNN'95-international conference on neural networks*. IEEE, 1995. 1942–1948.

Khan, Hamza, Saad Jamshed Abbasi, and Min Cheol Lee. "DPSO and inverse jacobian-based real-time inverse kinematics with trajectory tracking using integral SMC for teleoperation." *IEEE Access* 8 (2020): 159622–159638.

Kim, Nam-Ho, Dawn An, and Joo-Ho Choi. *Prognostics and health management of engineering systems*. Switzerland: Springer International Publishing, 2017.

Kircanski, Manja V., and Tatjana M. Petrovic. "Inverse kinematic solution for a 7 DOF robot with minimal computational complexity and singularity avoidance." *In Proceedings. 1991 IEEE International Conference on Robotics and Automation*. Sacramento, CA, USA: IEEE, 1991. 2664–2669.

Kumar, Ashwani, Vijay Kumar Banga, Darshan Kumar, and Thaweesak Yingthawornsuk. "Kinematics solution using metaheuristic algorithms." *In 2019 15th International Conference on Signal-Image Technology & Internet-Based Systems (SITIS)*. Sorrento, Italy: IEEE, 2019. pp. 505–510.

Kumar, K. Pavan, Mohan J. Murali, and D. Srikanth. "Generalized solution for inverse kinematics problem of a robot using hybrid genetic algorithms." *International Journal of Engineering & Technology* 7, no. 4.6 (2018).

Kumar, Rajesh, and Manoj Fozdar. "Optimal sizing of hybrid ship power system using variants of particle swarm optimization." *In 2017 Recent Developments in Control, Automation & Power Engineering (RDCAPE)*. Noida, India: IEEE, 2017. pp. 527–532.

Li, Chen, et al. "Inverse Kinematics Study for Intelligent Agriculture Robot Development via Differential Evolution Algorithm." *In 2021 International Conference on Computer, Control and Robotics (ICCCR)*. Shanghai, China: IEEE, 2021. pp. 37–41.

Marić, Filip, Matthew Giamou, Soroush Khoubyarian, Ivan Petrović, and Jonathan Kelly. "Inverse kinematics for serial kinematic chains via sum of squares optimization." *In 2020 IEEE International Conference on Robotics and Automation (ICRA).* Paris, France: IEEE, 2020. pp. 7101–7107.

Nearchou, Andreas C. "Solving the inverse kinematics problem of redundant robots operating in complex environments via a modified genetic algorithm." *Mechanism and machine theory* 33, no. 3 (1998): 273–292.

Nguyen, Thanh-Trung, Van-Huy Nguyen, and Xuan-Huong Nguyen. "Comparing the Results of Applying DE, PSO and Proposed Pro DE, Pro PSO Algorithms for Inverse Kinematics Problem of a 5-DOF Scara Robot." *In 2020 International Conference on Advanced Mechatronic Systems (ICAMechS).* Hanoi, Vietnam: IEEE, 2020. pp. 45–49.

Payal, Manju, Pooja Dixit, T.V.M. Sairam, and Nidhi Goyal. "Robotics, AI, and the IoT in defense systems." In *AI and IoT-Based Intelligent Automation in Robotics.* Hoboken, NJ: Wiley-Scrivener, 2021. pp. 109–128.

Poon, Joseph K., and Peter D. Lawrence. "Manipulator inverse kinematics based on joint functions." *In Proceedings. 1988 IEEE International Conference on Robotics and Automation.* Philadelphia, PA, USA: IEEE, 1988. pp. 669–674.

Qi, Yuhua, Jianan Wang, and Jiayuan Shan. "Aerial cooperative transporting and assembling control using multiple quadrotor–manipulator systems." *International Journal of Systems Science* 49, no. 3 (2018): 662–676.

Reis, João, Yuval Cohen, Nuno Melão, Joana Costa, and Diana Jorge. "High-tech defense industries: Developing autonomous intelligent systems." *Applied Sciences* 11, no. 11 (2021).

Reyes, Smith Vera, and Sixto Prado Gardini. "Inverse kinematics of manipulator robot using a PSO metaheuristic with adaptively exploration." *In 2019 IEEE XXVI International Conference on Electronics, Electrical Engineering and Computing (INTERCON).* Lima, Peru: IEEE, 2019. pp. 1–4.

Singh, Bharat, Rajesh Kumar, and Vinay Pratap Singh. "Reinforcement learning in robotic applications: A comprehensive survey." *Artificial Intelligence Review* 55, 2022. 945–990.

Tabandeh, Saleh, Christopher Clark, and William Melek. "A genetic algorithm approach to solve for multiple solutions of inverse kinematics using adaptive niching and clustering." *In 2006 IEEE International Conference on Evolutionary Computation.* Vancouver, BC: IEEE, 2006. pp. 1815–1822.

Umar, Abubakar, Zhanqun Shi, Wei Wang, and Zulfiqar Ibrahim Bibi Farouk. "A novel mutating pso based solution for inverse kinematic analysis of multi degree-of-freedom robot manipulators." *In 2019 IEEE International Conference on Artificial Intelligence and Computer Applications (ICAICA).* Dalian, China: IEEE, 2019. pp. 459–463.

Wang, Jianyu, Zhenzhou Shao, He Kang, Hongfa Zhao, Guoli Song, and Yong Guan. "Inverse kinematics of 6-DOF robot manipulator via analytic solution with conformal geometric algebra." *In 2018 IEEE International Conference on Robotics and Biomimetics (ROBIO).* Kuala Lumpur, Malaysia: IEEE, 2018. 2508–2513.

Wang, L-CT, and Chih-Cheng Chen. "A combined optimization method for solving the inverse kinematics problems of mechanical manipulators." *IEEE Transactions on Robotics and Automation* 7, no. 4 (1991): 489–499.

Wang, Xi Vincent, and Lihui Wang. "A literature survey of the robotic technologies during the COVID-19 pandemic." *Journal of Manufacturing Systems* 60 (2021): 823–836.

Wu, Fan, GuoQing Shi, and Sheng Qiang Wang. "Inverse kinematics solution of manipulator based on PSO-ELM." *In 2019 IEEE International Conference on Cybernetics and Intelligent Systems (CIS) and IEEE Conference on Robotics, Automation and Mechatronics (RAM)*. Bangkok, Thailand: IEEE, 2019. pp. 293–297.

Yotchon, Phanomphon, and Yutana Jewajinda. "Combining a differential evolution algorithm with cyclic coordinate descent for inverse kinematics of manipulator robot." *In 2021 3rd International Conference on Electronics Representation and Algorithm (ICERA)*. Yogyakarta, Indonesia: IEEE, 2021. pp. 35–40.

Chapter 11

Estimation of bearing remaining useful life using exponential degradation model and random forest algorithm

Pawan, Jeetesh Sharma, and Murari Lal Mittal

Malaviya National Institute of Technology, Jaipur, India

CONTENTS

11.1 INTRODUCTION

The rotating bearings are among the standard and essential equipment in mechanical systems, and it is critical to the machine's health. Bearing failure can result in a plant's abrupt shutdown, substantial losses, and possibly catastrophic disaster. As a result, precise bearing residual life estimation can increase production while reducing repair time and maintenance expenses. It also aids in the planning of maintenance strategies and the enhancement of overall system performance. These benefits of bearing remaining useful life (RUL) estimation have inspired several researchers' curiosity, resulting in the development of many approaches for bearing RUL estimation (Liu et al., 2018 & Heng et al., 2009).

Most methods for estimating a bearing's RUL are categorized as either data driven or model based. There are four types of approaches for estimating RUL in rotary machines: physical model-based approach, statistical model-based approach, AI-based approach, and some models based on hybrid strategies (Lei et al., 2018). Data-driven approaches use past measurement data, which is mainly collected from bearings via accelerometers (Soualhi et al., 2021, Lim & Mba, 2015, Medjaher et al., 2013 & Ahmad et al., 2018) and acoustic emission (AE) sensors (Elforjani, 2017, Elforjani, 2016, Elforjani & Shanbr, 2018, Elforjani et al., 2017). This historical data is studied to determine the bearing's health and understand its degradation through its lifetime. The gathered measurement data is utilized to evaluate the bearing's deterioration behavior, and that knowledge of behavior is subsequently used for predicting the bearing's RUL. Since it is easy to use computation as well as for data collecting through sensing techniques, the data-driven approach is one of the most widely employed of these strategies, especially for complex industrial systems (Kumar & Kumar, 2018, Heng et al., 2009, Ben Ali et al., 2015).

Machine learning algorithms–based models and historical degradation data collected from various sensors are used in data-driven techniques. Machine learning method associates bearing deterioration patterns with vibration signal characteristics or processed information. To determine a bearing's RUL, many data-driven methods have been developed. (Ning et al., 2018) constructed bearing health indicator using recurrent neural networks (RNN), and a particle filtering technique updated exponential model parameters for RUL prediction. Similarly, (Wang et al., 2020) developed a hybrid methodology for the prognosis of bearing RUL estimation. Different RVs are generated using a relevance vector machine (RVM), and the generated exponential degradation model fits RVs. Fréchet distance of the fitted curves is calculated for the optimal degradation curve for RUL prediction. Also, a hybrid prognostic technique based on a dynamic regression model was utilized to calculate the change rate of features and estimate the bearing's RUL (Ahmad et al., 2018 & Ahmad et al., 2019). This method uses the features rate change to find the start time of bearing deterioration and its failure threshold. While data-driven procedures rely on past measurement data, they can only be used on systems with such data. RUL prediction starts when a bearing shows symptoms of deterioration and deviates from its behavior in a constant state. The time to start prediction (TSP) is when the prediction process begins (Saxena et al., 2008). The correct identification of the TSP determines the accuracy of the RUL estimations. When a TSP is wrongly recognized, either critical information regarding early fault data is omitted or standard condition measurements are included in the RUL forecast. The TSP is frequently determined subjectively (Gebraeel, 2006, GEBRAEEL et al., 2005 & Si et al., 2013). Some authors, however, have offered other ways for determining the TSP, such as the $\mu + 3\sigma$ approach (Li et al., 2015 & Wang et al., 2016), when the system shows the longest constant time, engineering norms of ISO 10816, and machine-based procedures used statistical features of (Ginart et al., 2006 & Tran & Yang, 2012).

The authors (Lim & Mba, 2015) combined several switching Kalman filters by predicting distinct deterioration patterns of bearing to predict the RUL. The approach considers the unpredictability of prediction outcomes and is generalizable. However, the monotonical increment and degradation path for monitoring the condition using feature significantly impacts the bearing RUL prediction. Also, (Duan et al., 2018) applied a technique for data processing based on a cumulative transformation approach and employed it to train an extreme learning machine algorithm-based model to forecast the bearing deterioration pattern. A novel idea is used to change from prognostic work to classification work (Ben Ali et al., 2015) that employed a simplified fuzzy adaptive resonance theory map (SFAM) neural network to classify the different degradation stages of bearing. But the classification accuracy of this approach was low, and the bearing RUL estimation results depended on the number of iterations of the smoothing algorithm used in this work.

Similarly, (Yan et al., 2020) used SVM to categorize the bearing's deterioration stage and an approach-based smoothing to predict the best RUL at different phases of the deterioration stage of the bearing. The moment to begin prediction may be simply determined using the increased capabilities. Both ideas are new in prognostics of bearing, and robust algorithms can be used for classification and regression tasks. But still, there are some challenges present in current studies that require to be fixed in existing data-driven approaches. These significant challenges are as calculating the time period when the bearing enters in degradation phase, handling the random fluctuation of features used for prediction so that accuracy can be improved in RUL prediction, and calculating the failure point of bearings (Ahmad et al., 2018 & Yan et al., 2020).

In Table 11.4, the authors and their proposed methodology are discussed, which shows that data-driven methods are widely used to predict bearing RUL.

The key features of this chapter are as follows:

1. To extract dimensionless features from the historical bearing degradation dataset that can be used under the general operating conditions of bearings.
2. An exponential degradation model is used to track the degradation path of the bearing.
3. Prognostic work is transformed into classification work by random forest and prediction of RUL by the smoothing method.

The remainder of this document continues as follows: Section 11.2. provides the proposed approach for RUL estimation. Section 11.3 offers meticulous details of the PRONOSTIA dataset and information on the solution methodology, and analyzes the results obtained from numerical experiments. Section 11.4 provides a conclusion and future research directions.

11.2 THE PROPOSED RUL ESTIMATE APPROACH

The flow chart suggested approach is depicted in Figure 11.1, which contains two stages. During the training step, one run-to-failure bearing is utilized for training the RF classifier, which comprises specific information about the bearing's natural deterioration process. The trained RF classifies the processed features during the prediction step without fitting into multiple degradation stages. A smoothing algorithm has been used to predict the RUL of bearing, resulting in better prediction accuracy.

11.2.1 Features extraction

Authors have used numerous signal-processing techniques for feature extraction in the literature review. The root mean square (RMS) value in the time domain feature extraction from vibration acceleration signal is the most often employed characteristic for bearing prediction. It is favorably connected with the bearing deterioration trend (Wang et al., 2018). The spurious RMS variation significantly impacts the prognosis of bearing RUL (Medjaher et al., 2013). Figure 11.2 depicts such fluctuation in

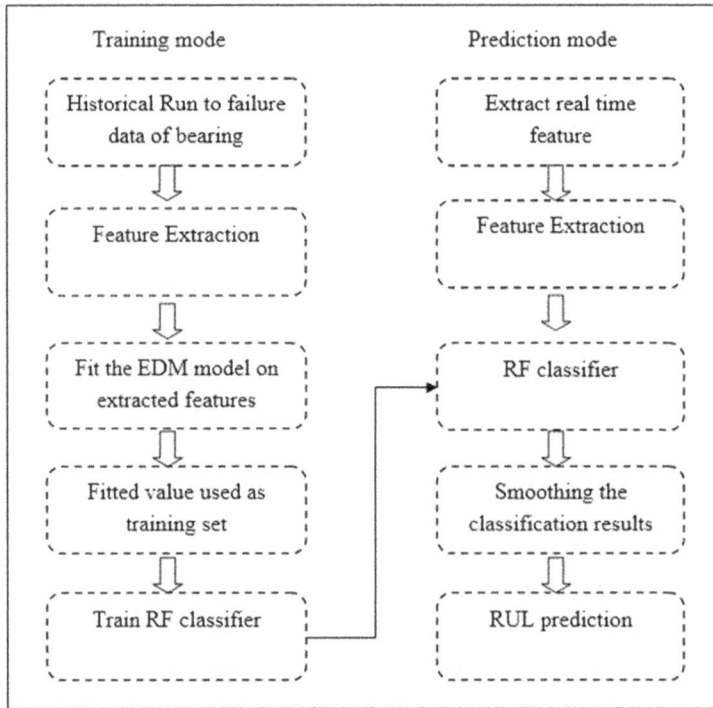

Figure 11.1 Flow chart of the proposed methodology.

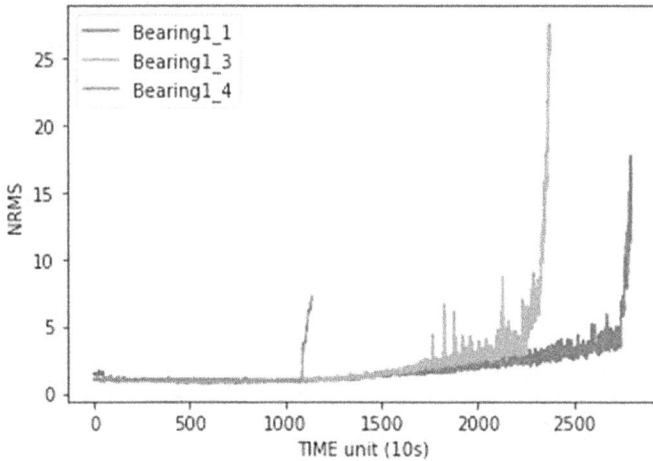

Figure 11.2 The behavior of NRMS value of different bearing under similar loading condition.

RMS values that results from the differing bearing run-to-failure historical dataset. Damage propagation on race causes RMS fluctuation. When minor fractures on the race began to appear and propagate, the RMS started to rise. The RMS decreased as the constant rolling contact smoothed the edges of tiny fractures. The RMS rose again as the bearing race's damaged region extended wider (WILLIAMS et al., 2001). Present sensor noise and other environmental vibrations from the machine cause fluctuations. Two features are proposed to capture the deterioration process to eliminate spurious instability and enhance RUL prediction ability.

11.2.2 Root mean square

ISO 2372 establishes an industrial norm for mechanical vibration: when the medium mechanical vibration signal's RMS value exceeds 2.0 to 2.2 g, the equipment is considered unsafe (Blake & Mitchel, 1972). The following equation may be used to calculate the RMS:

$$X_{rms} = \sqrt{\frac{1}{N} \sum_{i=1}^{N} (X(i))^2}$$

Figure 11.3 represents the behavior of the RMS value of bearing1_1.

The first used feature is the normalized root mean square (NRMS) and is calculated as follows:

$$NRMS(i) = \frac{RMS(i)}{RMS(normal)}$$

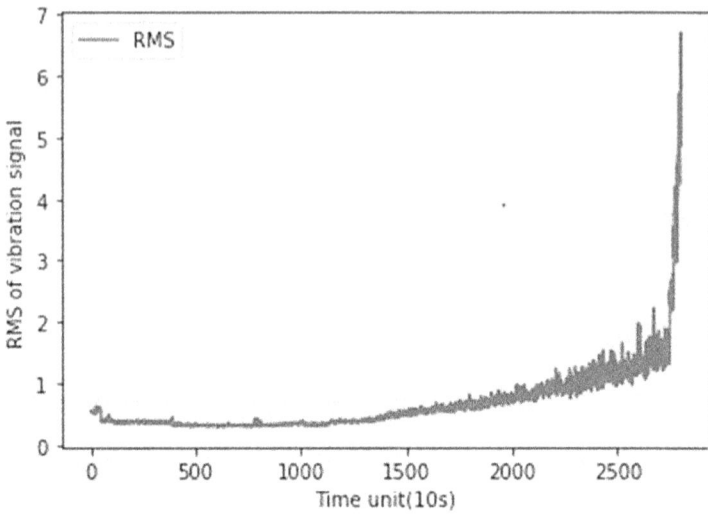

Figure 11.3 RMS behavior of bearing.

$$\mathrm{RMS(normal)} = \frac{1}{N}\sum_{i=1}^{N}\mathrm{RMS}(i)$$

During the steady operating stage of a bearing, RMS (normal) is the average mean square root. In this work, the data points range from 200 to 300 as the bearing's steady functioning stage. Individual bearing differences can be eliminated using this measurement, which causes a considerable impact on RMS and RUL prediction.

Even under identical testing conditions, as shown in Figure 11.2, the RMS of various bearings in the steady operation stage also varies with time. Individual changes in bearings have little effect on NRMS, which is simple to compute. It's also helpful in determining when to start predicting since NRMS intensifies the early bearing degeneration indicators and is sensitive to an impending bearing issue. The NRMS of the identical bearings, whose RMS values are presented in Figure 11.4. Because the NRMS characteristics remain practically constant before a bearing exhibits signs of degradation, NRMS = 1.1* of stable operation is assigned as one of the degeneration criteria in the beginning.

The smooth normalized root mean square (SNRMS) is the second suggested measurement based on NRMS, and it may be calculated as follows:

After the bearings' steady deterioration over time, degradation indicators can be used to indicate better variations. Modeling these stochastic changes in degradation indicators, which are not standard characteristics of bearing

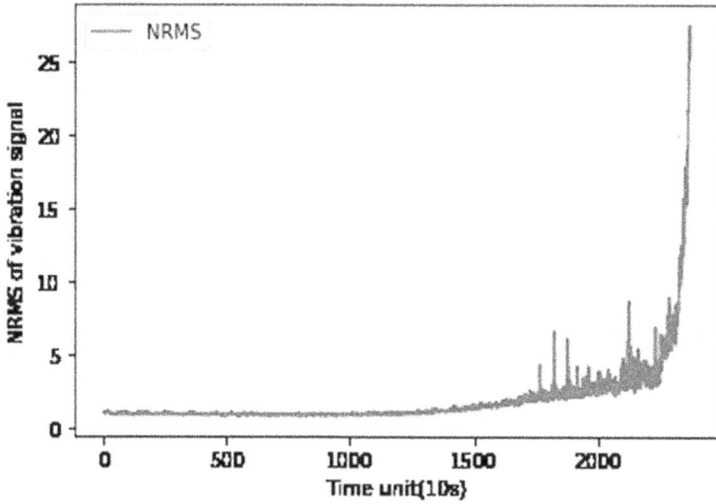

Figure 11.4 NRMS value of bearing.

deterioration, is problematic from a modeling standpoint. The normalized RMS is smoothed using linear rectification technology (LRT) to obtain SNRMS, which reduces the unpredictability of the deterioration signs; results are as follows:

$$X_{rms(i)} = \begin{cases} X_{rms(i)} \forall X_{rms(i)} \leq X_{rms(i)} \leq (1+\eta) X_{rms(i)} \\ X_{rms(i-1)} + \eta \;\; \forall X_{rms(i)} < X_{rms(i)} \vee X_{rms(i)} > (1+\eta) X_{rms(i)} \end{cases}$$

Here,
η = Growth rate

$$\eta = \frac{1}{N} \left| \sum_{i=1}^{N} X_{(rms)(i+1)} - X_{(rms)(i)} \right|$$

For smoothing the signal, the three-sigma rule is employed in the construction of instantaneous root mean square (IRRMS) shown in Figure 11.5, as the minimum sample size n should be 30 for sampling according to (Lehmann and Casella 1998). The size of the window (n = 30) is chosen to compute the SNRMS measurement because the sensitivity of SNRMS is directly connected to the classification results and smoothing techniques mentioned in this section.

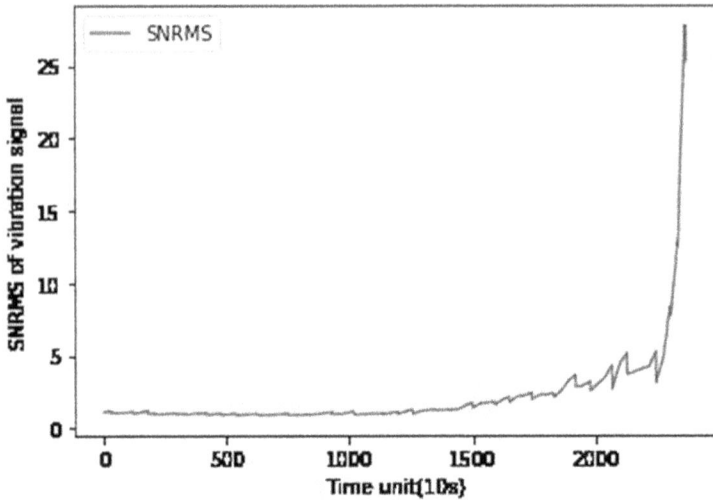

Figure 11.5 SNRMS value of bearing.

11.2.3 Feature fitting

The exponential degradation model is used for feature fitting because of its simplicity:

$$F(t) = a\exp(bt) + c$$

where F is the output value, i.e., fitted feature, t is the time, and a, b, c, and d are unknown parameters that can be determined using curve fitting methods. The nonlinear least square (Strutz, 2011) is employed to estimate these unknown parameters in this work.

Figure 11.6 shows the fitted value of the NRMS and SNRMS features, respectively, and the optimal parameters of EDM are given in Table 11.1. These fitted features do not have any random fluctuation and rise monotonically. These features are capable of defining the bearing deterioration process. After that, these fitted features are used to train RF. The fitted data represent well that the degeneration of the physical behavior of bearing is an irreversible process, given the mechanism of bearing degradation (Lei et al., 2018). As a result, the fitted features correlated with the degradation of bearing condition and accurately classified the stage of bearing degradation. These features support the benefits of employing fitted measures to train an RF classifier. During the training period, the fitted features are only utilized for training the RF classifier. The features (NRMS and SNRMS) are used as RF inputs during the prediction period without being fitted with EDM.

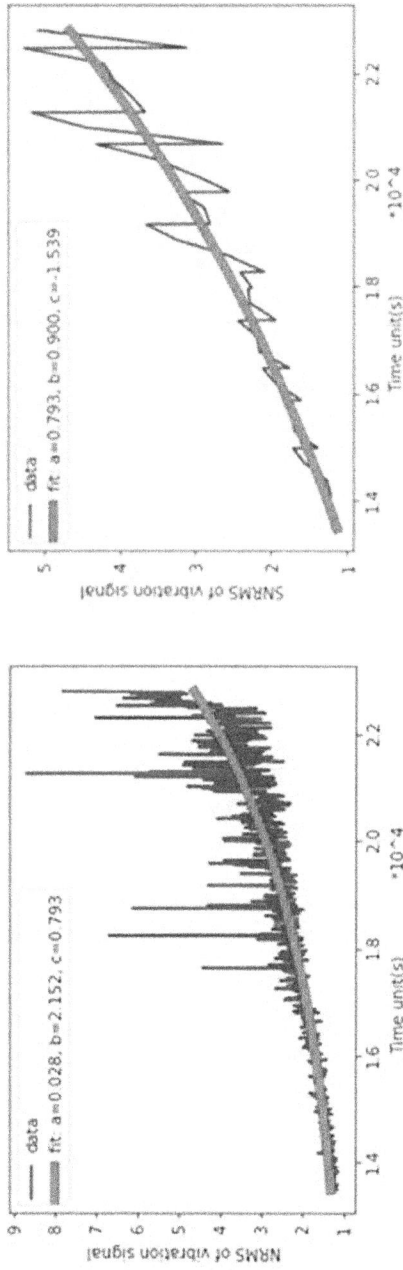

Figure 11.6 Fitted measurement of NRMS and SNRMS respectively by EDM model.

Table 11.1 Optimal values of parameters used in EDM model

	a	b	c
NRMS	0.028	2.152	0.793
SNRMS	0.793	0.900	−1.539

Table 11.2 RF parameters

	No. of estimators	Criterion	Random state
RF	10	entropy	zero

The fitting of the two measurements was taken to train the RF classifier, which is used to determine the stage of bearing deterioration. As a result, the procedures for using the RF classifier are suggested in this section (Table 11.2).

11.2.4 Training of random forest

It is required to train the RF algorithm on the bearing degradation fitted measurement before the prediction of bearing RUL; the details of training and testing are given as follows:

Input to RF: First, the fitted features are scaled linearly between 0 and 1 and, after that, used as input to train the RF. Since scaling has the advantage of preventing more prominent features and numeric values from dominating features over the smaller numeric values feature and computational problems throughout the process. Two features are used as input to train the RF classifier.

The output of RF: The output data is classified into the classes (L), where class 1 shows the healthy state of the bearing followed by the remaining classes of different stages of bearing deterioration. In actual applications, the failure threshold is used to determine the number of classes. In this work, L=6 is classified by the RF classifier. It's worth noting that L is fine-tuned empirically utilizing a variety of common run-to-failure history data.

In Figure 11.7, the phrase "Class 2, Deterioration 30 percent" implies that if a sample point is in this stage (Class 2), that is, the proportion of time from the point of degradation beginning to this sample point in the complete degradation development period from 30% to 60%, then it has a degradation index of 30% (Table 11.3). Construction of the degradation index from large to small helps in reducing the influence of variation on RUL forecast accuracy and increases the model's flexibility. Several representative whole-life vibration monitoring data are also used to tune the deterioration

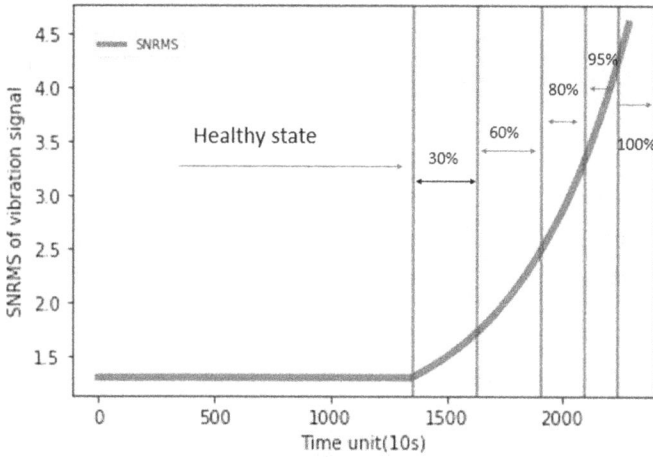

Figure 11.7 Outputs of RF classifier.

Table 11.3 Division of the state of bearing1_3

Class	Time (s)	% degradation
0	Start–13,500	Healthy state
1	13,510–16,290	0–30%
2	16,300–19,090	30–60%
3	19,100–20,950	60–80%
4	20,960–22,340	80–95%
5	22,350–22,820	95–100%

index in different classes empirically. For different degradation stages, the initial value ($SNRMS_{initial}$) and the threshold value ($SNRMS_{final}$) for each category are given in Table 11.4.

Training of RF used the fitted bearing measurements as inputs and outputs, which comprise specific deterioration process information and its associated class.

During the prediction stage of RF, the measurements are provided as inputs to the trained RF after they are obtained from the bearing without fitting it. Each input vector will be assigned to a class that indicates the bearing's deterioration stage. As a result, a bearing deterioration stage evaluation is carried out, which is beneficial for bearing RUL estimation. If any data point falls beyond the threshold failure value or falls outside the defined classes, this value is automatically classified as the last class. This condition often arises in practice while evaluating the maintenance approach to avoid an abrupt plant closure.

Table 11.4 Initial and final values for each class

Class ⇨	1	2	3	4	5
SNRMS$_{initial}$	1.19	1.98	3.50	4.12	4.74
SNRMS$_{final}$	1.98	3.50	4.12	4.74	5.07
Class ⇨	1	2	3	4	5
SNRMS$_{initial}$	1.19	1.98	3.50	4.12	4.74
SNRMS$_{final}$	1.98	3.50	4.12	4.74	5.07

11.2.5 RUL prediction

Once the class of each input vector is obtained by a classification algorithm, a preliminary estimate RUL of bearings is accepted for the bearing's deterioration stage. However, monitoring the deterioration process in the same deterioration phase requires another approach to get more precise RUL predictions of bearings. Therefore, a smoothing prediction algorithm is used for better RUL prediction. The suggested smoothing algorithm's primary goal is to compute the average of each two succeeding classes.

11.3 EXPERIMENTAL RESULT AND DISCUSSION

The following section contains the development of the proposed methodology.

11.3.1 Experimental data

Researchers utilized the PRONOSTIA platform to perform accelerated deterioration experiments on ball bearings for a few hours, allowing them to analyze bearing degradation across their entire operating life. There are time frames of realistically feasible duration, from typical healthy states to failure due to numerous bearing problems. Furthermore, as compared to degradation testing conducted over several days, the data collected in these accelerated degradation tests were substantially insufficient in amount.

At the commencement of the testing, all of the bearings used to collect the accelerated deterioration test findings were normal and had no bearing issues. During the testing, these bearings usually deteriorated, and all of them failed in the end. The rings, balls, and cage of each failed bearing had defects. The threshold criterion of the vibration acceleration signal exceeded 20 g. Figure 11.9 depicts the PRONOSTIA platform (Medjaher et al., 2013). It has three primary components: an AC motor in the rotating portion, a gearbox or speed reducer, and two shafts. A loading component consists of a pneumatic control valve and a force transmission mechanism applied radial load on the spinning bearings, causing them to deteriorate faster. The measuring component consists of sensors that identify the current operating conditions, such as force sensors, speed sensors, torque meters, sensors that measure bearing degradation, such as vibration sensors or accelerometers, and a thermocouple.

Power is provided to the secondary shaft through a gearbox from the asynchronous AC motor, which generates about 250 W (designated as a speed reducer in Figure 11.3). The gearbox or speed reducer keeps the secondary shaft speed below 2,000 rev/min (revolutions per minute) while providing the rated torque. The motor is running at its rated speed of 2,830 r/min, as indicated in Table 11.5. The rotating motion of the secondary shaft is transmitted to the bearings through a compliant and robust shaft connection, as shown in Figure 11.3. The loading portion of the PRONOSTIA platform is seen in detail in Figure 11.9. The pneumatic jack initially provides the force or load and is then increased by a lever arm, as shown in Figure 11.9. The increased force is then transferred indirectly to the test bearing's outer ring. As previously indicated, the measurement section comprises two sets of sensors, one for recognizing operating conditions and the other for assessing bearing degradation. The working conditions are determined by the radial load given to the test bearings, the speed of rotation of the secondary shaft, and the torque imparted to the bearing. Figure 11.8 shows how a resistance temperature detector (RTD) platinum probe (PT100) placed into a hole near the bearing's outer ring detects the bearing temperature during operation. The RTD sensor is sampled ten times per second. As a result, bearing

Table 11.5 Details about the bearing's operational conditions as well as the results of the accelerated deterioration test

Operating condition of bearing			
	Loading condition 1	Loading condition 2	Loading condition 3
Radial loading (N)	4,000	4,200	5,000
Speed (rev/min)	1,800	1,650	1,500
Learning set	Bearing1_1	Bearing2_1	Bearing3_1
	Bearing1_2	Bearing2_2	Bearing3_2
	Bearing1_3	Bearing2_3	Bearing3_3
Test set	Bearing1_4	Bearing2_4	
	Bearing1_5	Bearing2_5	
	Bearing1_6	Bearing2_6	
	Bearing1_7	Bearing2_7	

```
Initialization
Fix the number of smoothing iterations si
Set the start smoothing time sst
Repeat
Initialize the smoothing step ss=0
Repeat
Compute the smoothed RUL by
RUL(sts)=RUL(sts)+RUL(sts−1)
            2
Increment the smoothing by ss=ss+1
Until (ss=si)
Select the next point to inspect it
Until (the end of the experiment)
```

Figure 11.8 Smoothing prediction algorithm (Ben Ali et al., 2015).

temperature is measured every 1/10th of a second, resulting in 600 temperature measurements each minute.

These three measurements are made at a 100 Hz frequency and are used
to determine the operating conditions. On the PRONOSTIA platform, two
measurements may be collected to characterize bearing deterioration: vibration and temperature data. However, vibration data is only used in this
study to assess bearing degradation and estimate RUL. To detect bearing
vibration along both the horizontal and vertical axes, two accelerometers
are positioned radially on the housing of the test bearing at 90° to each
other, as shown in Figure 11.10. At a sample rate of 25.6 kHz, the vibration
acceleration is recorded for 1/10th of a second every 10 seconds.
Consequently, each vibration signal recording has 2,560 samples, with 10
seconds between each pair of recordings.

In this section, the results of the classification and the prediction of RUL
are discussed. Since bearing, RUL predictions should be established using the
same failure criteria. The failure threshold in this study is set at NRMS = 7,
which industry standards can determine based on the individual application.
An assumption is made that the deterioration is to be progressive rather than
abrupt. Because the failure threshold is dimensionless, it is unaffected by the
magnitude of bearing measurements or individual bearing variations. RUL
prediction using fitted outcomes is intended to be more precise than dimension measurements. Therefore, In the PRONOSTIA dataset, bearing 1 of testing 1, bearing 3 of testing 1, and bearing 4 of testing 1 are chosen to validate
the proposed methods. The selected bearing 4 of testing 1 in the PRONOSTIA
dataset has a sudden failure mode. This can be considered as the proposed
method performs better in such fluctuations.

11.3.2 RUL prediction using the proposed approach

It is observed that after having the same bearing type and operating condition, the process of degradation, failure duration, and vibration signal are
different for different bearings. These are the challenges in predicting the
RUL of a bearing. This approach is applied to some selected bearings of

Figure 11.9 Bearing degradation test on PRONOSTIA platform (Medjaher et al., 2013).

Reliability Engineering and System

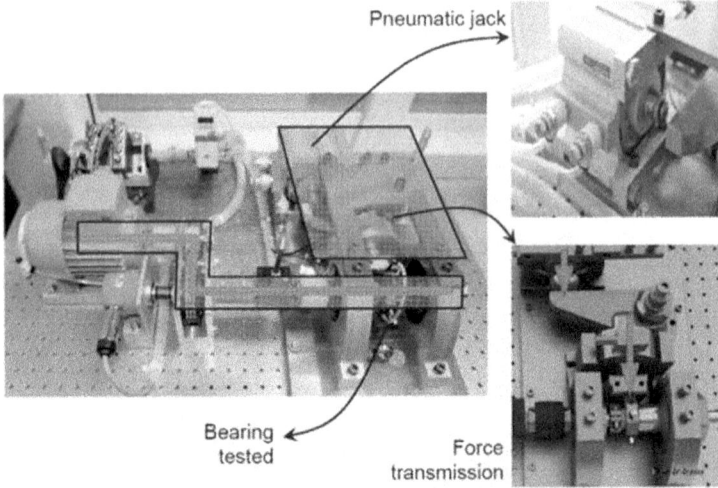

Figure 11.10 Loading parts detailed figure (Medjaher et al., 2013).

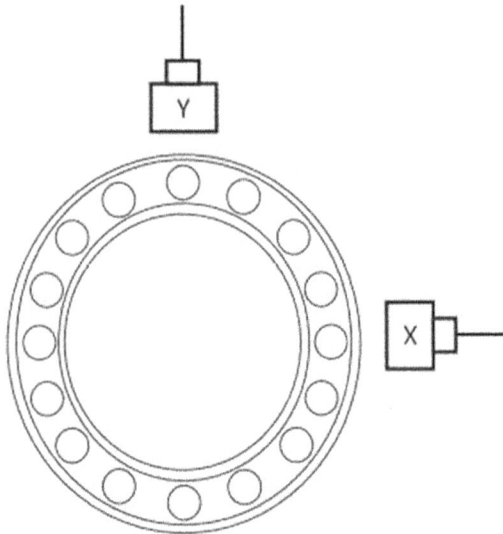

Figure 11.11 The accelerometers are positioned on the horizontal X-axis and the vertical Y-axis of the bearing housing in this schematic diagram (Ahmad et al., 2019).

the PRONOSTIA dataset, as mentioned earlier. The RF's algorithm inputs are NRMS and SNRMS data that have been linearly scaled between 0 and 1 before inputting into the classification algorithm. The following is the definition of the input vector I:

$$I = \left[NRMS(i), SNRMS(i) \right],$$

where I represents the current inspection value time.

The RF's outputs are divided into six categories, each representing a distinct phase of bearing deterioration (L = 6). First, the fitted measurements (NRMS, SNRMS) of bearing 3 of testing 1 are used from the PRONOSTIA dataset for the training of the RF classifier. The optimal value of the parameters of the RF classifier is shown in Table 11.2. The trained RF algorithm is applied to the actual dataset of bearing 3 in testing 1. Each data point inspected on trained RF achieved a class accuracy of 98%. These classes represent different values of the degradation index, as shown in Figure 11.7, which are 30%, 30%, 20%,15%, and 5%, respectively. Table 11.6 shows the classification accuracy of several classes. The smoothing algorithm is used at different phases of bearing degradation to provide a more accurate RUL prediction. The basic concept of the smoothing algorithm is to follow the bearing's deterioration path. The average in the appropriate class is used as a smooth data point. Besides the advantage of the smoothing algorithm, it is also affected by the number of averaging iterations, and that is also a limitation of this technique. The averaging iterations for smoothing si are set to six (si = 8) in this work. It is essential to decide the initial value ($SNRMS_{initial}$) and final value ($SNRMS_{final}$) for each class of the degradation stage that is given in Table 11.4. These values are calculated based on the degradation index of bearing 3 of testing 1, as shown in Figure 11.12.

The tested bearing dataset is used for RUL calculation and is discussed as follows: the projected life percentage is 40% when the failure time is 2,282 (unit of time) and is noted at a time of 1,729 (unit of time). The criterion for

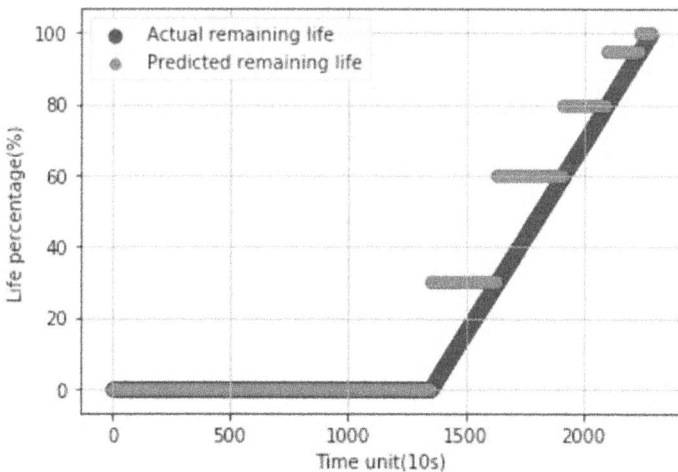

Figure 11.12 Classification result of bearing1_3.

Table 11.6 Classification accuracy at differing stages

Class	0	1	2	3	4	5
No of points	1350	279	279	185	139	48
Classified points	1347	267	271	184	135	46
Classification accuracy	99.77%	95.69%	97.13%	99.45%	97.12%	95.83%

predicted failure time after degeneration beginning is defined as NRMS = 1.1*NRMS, and at the time of 1,350 seconds the RUL would be (1,729–1,350)/40% = 948 (10 min). Based on the calculation, the RUL of the tested bearing would be (948 + 1,350)–1850 = 448 (10 min). The predicted failure time will be 1,850 + 448 = 2,298 (10 min). Relative accuracy (RA) of prediction will be an absolute value of 1-((2,282–2,298)/2,282), i.e., 99.29%.

11.3.2.1 Relative accuracy

RA is a measure of prediction results, i.e., how closely prediction results are, and this metric is widely used and given as follows:

$$RA_\lambda = 1 - \frac{\left| r_*(t_\lambda) - r(t_\lambda) \right|}{r_*(t_\lambda)}$$

where $t_\lambda = t_p + \lambda(\text{EOL} - t_p)$

The estimated value of RUL is given by an actual value of RUL at any given point in time (Saxena et al., 2021).

11.3.2.2 Cumulative relative accuracy

RA provides data at a given point in time. RA can be assessed in instances to produce an aggregate accuracy level or the cumulative relative accuracy (CRA) to determine the algorithm's overall behavior (CRA) (Saxena et al., 2021).

$$CRA_\lambda = \frac{1}{|l_\lambda|} \sum_{i \in l_\lambda} w(r(i)) RA_\lambda$$

Here,

At all-time indices, w(r(i)) is a weight factor as a function of RUL.

When a prediction is made, l_λ, it is the collection of all-time indexes.

Because strong performance near end of life (EoL) is essential for condition-based decision-making, it may be beneficial to assign more weight to RA assessed at times closer to EoL (Table 11.7).

Table 11.7 Classification results of the proposed method

Test Bearing	Classification accuracy (%)
Bearing1_1	88.54
Bearing1_3	98.23
Bearing1_4	95.86

The prediction results of the proposed method are shown in Figures 11.13, 11.14 and compared with other approaches used for bearing RUL prediction, shown in Table 11.8.

11.4 CONCLUSION

Estimation of bearing useful life method is proposed using EDM and random forest that can be generalized and used for accurate RUL prediction. Two features are used to show the vibration intensity of bearings in comparison to the typical vibration value. They can minimize individual bearing variations, enhance sensitivity to bearing emerging defects, and decrease variability. Furthermore, they aid in determining when to begin the prediction of RUL and establishing dimensionless failure thresholds. A random forest classifier is used on fitted measurements of the EDM model, which gives higher classification accuracy. A smoothing algorithm is used for better RUL prediction under different degradation stages. For validation, a

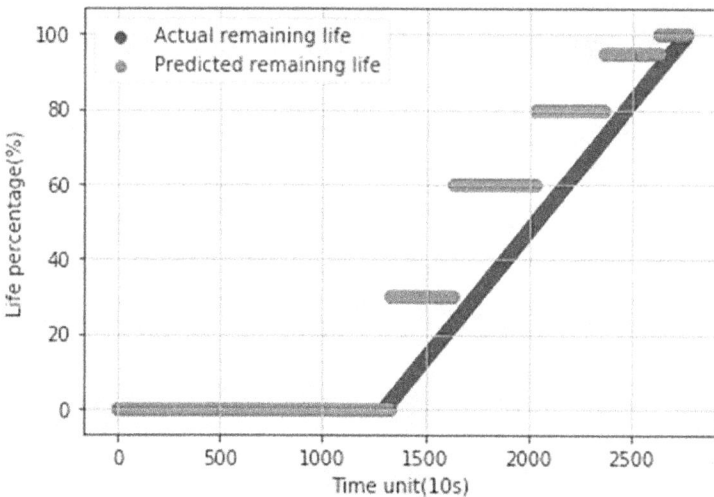

Figure 11.13 Classification result of bearing1_1.

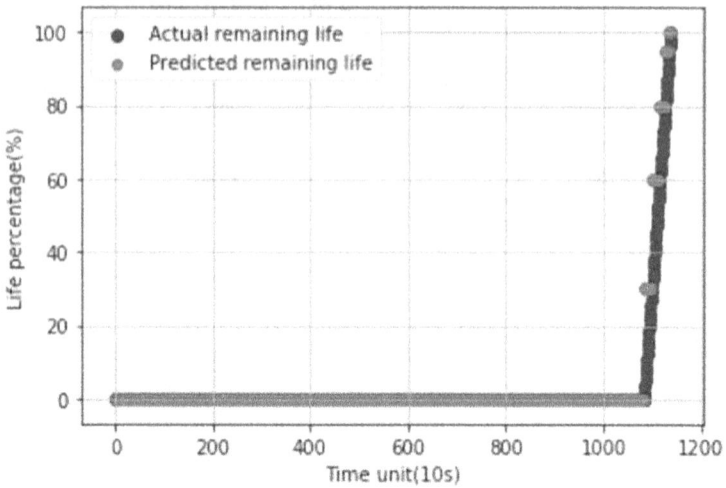

Figure 11.14 Classification result of bearing1_4.

Table 11.8 Comparison of CRA scores of RUL results

Test bearing	Exponential model	Improved exponential model	Hybrid SVM model	EDM with RF
Bearing1_1	0.7111	0.8696	0.8787	0.8854
Bearing1_3	0.542	0.8712	0.9421	0.9324
Bearing1_4	0.7463	0.9324	0.5562	0.5743

study based on SVM is compared with the proposed method. This suggested approach's performance was evaluated by computing CRA and comparing it to existing techniques for RUL prediction on the PRONOSTIA dataset. The outcomes found that the proposed approach performs well, particularly on the bearing during the gradual deterioration phase.

To increase the prediction accuracy, some better feature extraction methods can extract more accurate degradation trends of bearing from raw vibration data. This improves the robustness of the proposed technique by separating bearing fault signals and suppressing the signal from loud heavy sounds and other background noise signals. The classification technique used on the bearing deterioration stage, such as an extreme learning machine, XGboost, can be used in further study to improve accuracy and generalization.

REFERENCES

Ahmad, W., Khan, S., Islam, M., & Kim, J. (2019). A reliable technique for remaining useful life estimation of rolling element bearings using dynamic regression models. *Reliability Engineering & System Safety*, 184, 67–76. https://doi.org/10.1016/j.ress.2018.02.003

Ahmad, W., Khan, S., & Kim, J. (2018). A hybrid prognostics technique for rolling element bearings using adaptive predictive models. *IEEE Transactions On Industrial Electronics*, 65(2), 1577–1584. https://doi.org/10.1109/tie.2017.2733487

Ben Ali, J., Chebel-Morello, B., Saidi, L., Malinowski, S., & Fnaiech, F. (2015). Accurate bearing remaining useful life prediction based on Weibull distribution and artificial neural network. *Mechanical Systems And Signal Processing*, 56–57.

M.P. Blake, W.S. Mitchel. (1972). *Vibration and acoustic measurement*. Spartan Books, New York.

Duan, L., Zhao, F., Wang, J., Wang, N., & Zhang, J. (2018). An integrated cumulative transformation and feature fusion approach for bearing degradation prognostics. *Shock And Vibration*, 2018, 1–15. https://doi.org/10.1155/2018/9067184

Elforjani, M. (2016). Estimation of remaining useful life of slow speed bearings using acoustic emission signals. *Journal of Nondestructive Evaluation*, 35(4). https://doi.org/10.1007/s10921-016-0378-0

Elforjani, M. (2017). Diagnosis and prognosis of slow speed bearing behavior under grease starvation condition. *Structural Health Monitoring*, 17(3), 532–548. https://doi.org/10.1177/1475921717704620

Elforjani, M., & Shanbr, S. (2018). Prognosis of bearing acoustic emission signals using supervised machine learning. *IEEE Transactions On Industrial Electronics*, 65(7), 5864–5871. https://doi.org/10.1109/tie.2017.2767551

Elforjani, M., Shanbr, S., & Bechhoefer, E. (2017). Detection of faulty high speed wind turbine bearing using signal intensity estimator technique. *Wind Energy*, 21(1), 53–69. https://doi.org/10.1002/we.2144

Gebraeel, N. (2006). Sensory-updated residual life distributions for components with exponential degradation patterns. *IEEE Transactions On Automation Science And Engineering*, 3(4), 382–393. https://doi.org/10.1109/tase.2006.876609

Gebraeel, N., Lawley, M., Li, R., & Ryan, J. (2005). Residual-life distributions from component degradation signals: A Bayesian approach. *IIE Transactions*, 37(6), 543–557. https://doi.org/10.1080/07408170590929018

Ginart, A., Barlas, I., Goldin, J., & Dorrity, J. (2006). Automated feature selection for embeddable prognostic and health monitoring (PHM) architectures. *2006 IEEE Autotestcon*. https://doi.org/10.1109/autest.2006.283625

Heng, A., Zhang, S., Tan, A., & Mathew, J. (2009). Rotating machinery prognostics: State of the art, challenges and opportunities. *Mechanical Systems And Signal Processing*, 23(3), 724–739. https://doi.org/10.1016/j.ymssp.2008.06.009

Kumar, A., & Kumar, R. (2018). Role of signal processing, modeling and decision making in the diagnosis of rolling element bearing defect: A review. *Journal of Nondestructive Evaluation*, 38(1). https://doi.org/10.1007/s10921-018-0543-8

Lehmann, E. L., & George Casella. (1998). Theory of Point Estimation. *Texts in Statistics*. https://doi.org/10.1007/b6987310.1007/978-0-387-22728-3

Lei, Y., Li, N., Guo, L., Li, N., Yan, T., & Lin, J. (2018). Machinery health prognostics: A systematic review from data acquisition to RUL prediction. *Mechanical*

Systems And Signal Processing, 104, 799–834. https://doi.org/10.1016/j.ymssp.
2017.11.016

Li, N., Lei, Y., Lin, J., & Ding, S. (2015). An improved exponential model for predicting remaining useful life of rolling element bearings. *IEEE Transactions On Industrial Electronics*, 62(12), 7762–7773. https://doi.org/10.1109/tie.2015.2455055

Lim, C., & Mba, D. (2015). Switching Kalman filter for failure prognostic. *Mechanical Systems And Signal Processing*,52–53, 426–435. https://doi.org/10.1016/j.ymssp.2014.08.006

Liu, H., Mo, Z., Zhang, H., Zeng, X., Wang, J., & Miao, Q. (2018). Investigation on rolling bearing remaining useful life prediction: A review. *2018 Prognostics And System Health Management Conference (PHM-Chongqing)*. https://doi.org/10.1109/phm-chongqing.2018.00175

Medjaher, K., Zerhouni, N., & Baklouti, J. (2013). Data-driven prognostics based on health indicator construction: Application to PRONOSTIA's data. *2013 European Control Conference (ECC)*. https://doi.org/10.23919/ecc.2013.6669223

Ning, Y., Wang, G., Yu, J., & Jiang, H. (2018). A feature selection algorithm based on variable correlation and time correlation for predicting remaining useful life of equipment using RNN. *2018 Condition Monitoring And Diagnosis (CMD)*. https://doi.org/10.1109/cmd.2018.8535843

Saxena, A., Celaya, J., Balaban, E., Goebel, K., Saha, B., Saha, S., & Schwabacher, M. (2008). Metrics for evaluating performance of prognostic techniques. *2008 International Conference On Prognostics And Health Management*. https://doi.org/10.1109/phm.2008.4711436

Saxena, A., Celaya, J., Saha, B., Saha, S., & Goebel, K. (2021). Metrics for offline evaluation of prognostic performance. *International Journal of Prognostics and Health Management*, 1(1). https://doi.org/10.36001/ijphm.2010.v1i1.1336

Si, X., Wang, W., Chen, M., Hu, C., & Zhou, D. (2013). A degradation path-dependent approach for remaining useful life estimation with an exact and closed-form solution. *European Journal of Operational Research*, 226(1), 53–66. https://doi.org/10.1016/j.ejor.2012.10.030

Soualhi, M., El Koujok, M., Nguyen, K., Medjaher, K., Ragab, A., & Ghezzaz, H. et al. (2021). Adaptive prognostics in a controlled energy conversion process based on long- and short-term predictors. *Applied Energy*, 283, 116049. https://doi.org/10.1016/j.apenergy.2020.116049

T. Strutz, *Data Fitting and Uncertainty: A Practical Introduction to Weighted Least Squares and Beyond*, 1st ed. Wiesbaden, Germany: Vieweg and Teubner, 2011.

Tran, V., & Yang, B. (2012). An intelligent condition-based maintenance platform for rotating machinery. *Expert Systems with Applications*, 39(3), 2977–2988. https://doi.org/10.1016/j.eswa.2011.08.159

Wang, B., Lei, Y., Li, N., & Li, N. (2020). A hybrid prognostics approach for estimating remaining useful life of rolling element bearings. *IEEE Transactions On Reliability*, 69(1), 401–412. https://doi.org/10.1109/tr.2018.2882682

Wang, D., Tsui, K., & Miao, Q. (2018). Prognostics and health management: a review of vibration based bearing and gear health indicators. *IEEE Access*, 6, 665–676. https://doi.org/10.1109/access.2017.2774261

Wang, Y., Peng, Y., Zi, Y., Jin, X., & Tsui, K. (2016). A two-stage data-driven-based prognostic approach for bearing degradation problem. *IEEE Transactions On Industrial Informatics*, 12(3), 924–932. https://doi.org/10.1109/tii.2016.2535368

Williams, T., Ribadeneira, X., Billington, S., & Kurfess, T. (2001). Rolling element bearing diagnostics in run-to-failure lifetime testing. *Mechanical Systems And Signal Processing*, 15(5), 979–993. https://doi.org/10.1006/mssp.2001.1418

Yan, M., Wang, X., Wang, B., Chang, M., & Muhammad, I. (2020). Bearing remaining useful life prediction using support vector machine and hybrid degradation tracking model. *ISA Transactions*, 98, 471–482. https://doi.org/10.1016/j.isatra.2019.08.058

Chapter 12

Machine learning-based predictive maintenance for diagnostics and prognostics of engineering systems

Ramnath Prabhu Bam, Rajesh S. Prabhu Gaonkar, and Clint Pazhayidam George
Indian Institute of Technology Goa, India

CONTENTS

12.1 INTRODUCTION AND OVERVIEW

In simple terms, to carry out maintenance is to perform the activities required to keep machinery, equipment, or infrastructure in good condition. It involves repairs, replacements, servicing, or simple, functional checks. Maintenance aims to reduce the number of failures and avoid breakdowns that lead to unnecessary downtime. It also intends to ensure better product quality by preventing malfunctioning of the system. A safer work environment is also an outcome of an effective maintenance program, thereby preventing any injuries and fatal accidents by avoiding failures.

Once any machine or equipment starts functioning, we expect it to run under normal conditions, i.e., without fault. However, some faults might get initiated in the system over a while. Equipment health will degrade quickly until failure if we do not address faults through proper maintenance. However, with proper maintenance, the degradation in equipment health can be slowed down, thereby delaying the failure and allowing an increased functional time, "Δt." Figure 12.1 illustrates the effect of maintenance on equipment health over its functional time. A good maintenance strategy will directly improve the assets availability, thereby catalyzing the production process.

Maintenance Strategies: Maintenance is divided broadly into different categories; see Figure 12.2. Reactive maintenance (RM) works on the principle of "fix it when it breaks." It is a strategy that evokes a corrective action when something breaks down, and is referred to as corrective maintenance. Even though this strategy sounds unplanned, there is no initial cost and less requirement of planning time (Ozgur Unluakın et al., 2019). When the components are not critical to the functioning of the system or their failure would not be a costly affair to the organization, it is advisable to adopt an RM program. On the other hand, this scheme may result in a significant downtime if the failures are critical and complicated, thus affecting the business.

Preventive maintenance (PvM) is a schedule-based maintenance of equipment and machinery to keep them running and avoid failures. Its goal is to reduce the probability of failures, and is sometimes referred to as predetermined maintenance. We perform PvM when the equipment or asset is still

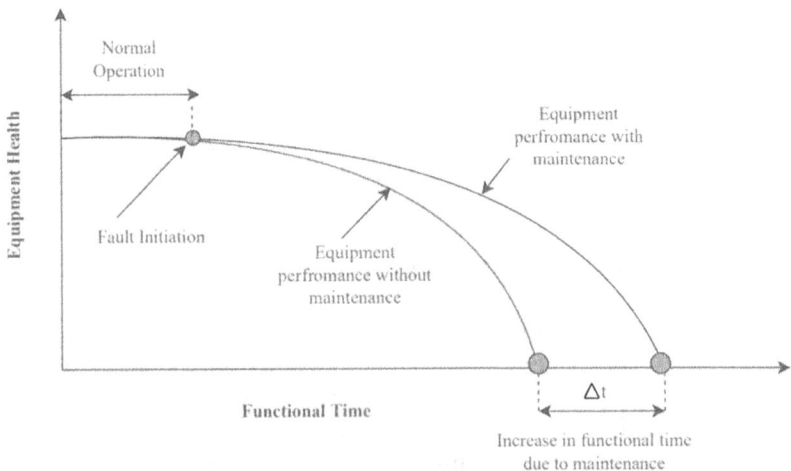

Figure 12.1 Degradation rate of equipment health with and without maintenance. Over a period of time, any equipment will develop some fault, which leads to faster health degradation. Equipment failure happens when equipment health is zero. With proper maintenance, this failure can be delayed, thereby providing a larger functional time.

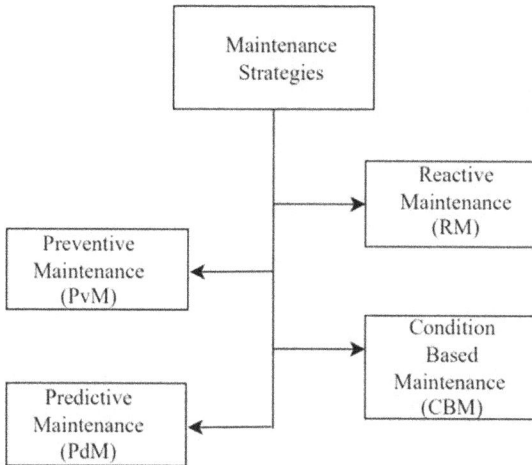

Figure 12.2 General categorization of maintenance strategies.

functional, and we can classify it into time-based PvM and usage-based PvM (Zheng and Zhou, 2021). Time-based PvM carries out the maintenance activity at regular time intervals. For example, we may replace oil filters after every three months. However, in usage-based PvM maintenance activity is scheduled based on specific usage terms. For example, the recommendation of servicing the vehicle after a few thousand kilometers is a usage-based PvM. Compared with RM, this scheme fares better as it does not wait until a failure happens and thereby significant production halts. However, it may bring a disadvantage when we do schedule-based maintenance targeting brand new components or assets, with rare chances of failure. It would lead to wasteful utilization of time and resources.

In summary, we wish to have a maintenance strategy that neither waits until breakdown happens nor demands any unnecessary maintenance. This requirement leads us to condition-based maintenance (CBM) (Duan et al., 2020). Its central idea is to raise concern when observing something abnormal. A CBM expert will then assess the condition of the equipment or machinery and develop a course of action. We may consider the analogy of human beings visiting physicians upon observing early signs of discomfort. In our discussion, we denote it as diagnostics and prognostics in this chapter. There are various tools and techniques available to perform CBM effectively (see Figure 12.3). One popular technique among them is vibration analysis (VA).

Although CBM looks promising, diagnostics and prognostics based on CBM require detecting and reporting first signs of abnormal behavior. If this goes undetected, then the fault or the abnormality will aggravate until detected. It focuses on a method that involves detecting the presence of faults or predicting their occurrence in the near future in any machine or equipment. This is achieved by continuously monitoring the data received

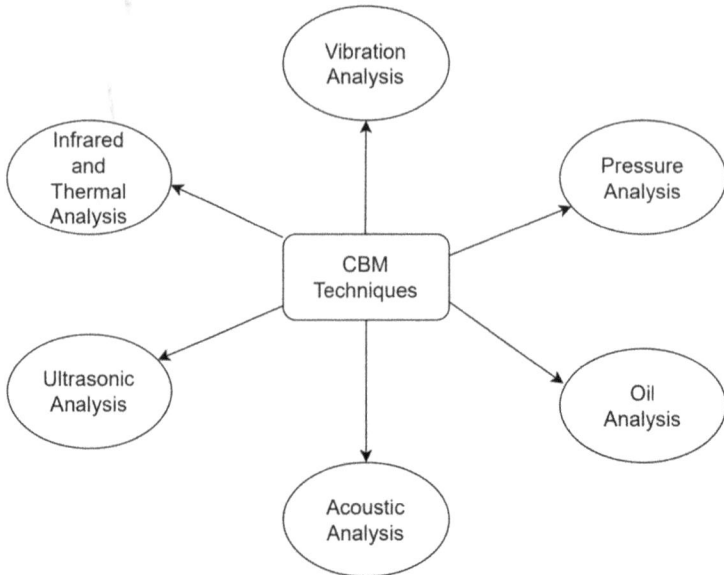

Figure 12.3 Popular techniques for condition-based monitoring. All these techniques help in determining whether the equipment is in normal working condition or there is some fault in it.

from such machines or equipment in the form of temperature, pressure, vibrations, acoustic signals, etc. Such a maintenance approach is called predictive maintenance (PdM) (Bhat et al., 2021). The main principle behind the PdM approach is to continuously monitor the system log data and raise the alarm as soon as the first sign of abnormality is detected. We aim to ensure the early detection of irregularity, which can avoid potential failures. PdM provides benefits like decreasing the cost of maintenance, lowering the unplanned downtime, and extending the equipment life. Thus, PdM has become a widely used diagnostic and prognostics technique in the modern-day industry.

At the nucleus, the PdM process involves data collection, preprocessing, and decision-making (see Figure 12.4). PdM uses data recorded from various sensors mounted on the equipment. The data can be in different signals such as vibration, acoustics, or measurements obtained from other sensors like temperature, pressure, speed, etc. The choice of data depends on the application and the problem to be solved.

PdM has grown in popularity in recent years also due to the rise of the Internet of things (IoT), artificial intelligence (AI), and machine learning (ML). The increasing popularity of PdM is also boosted by various businesses that have claimed to achieve superior results with its implementation.

Chapter organization: Section 12.2 elaborates on the fundamentals of diagnostics and prognostics based on PdM. Section 12.3 explains the basic

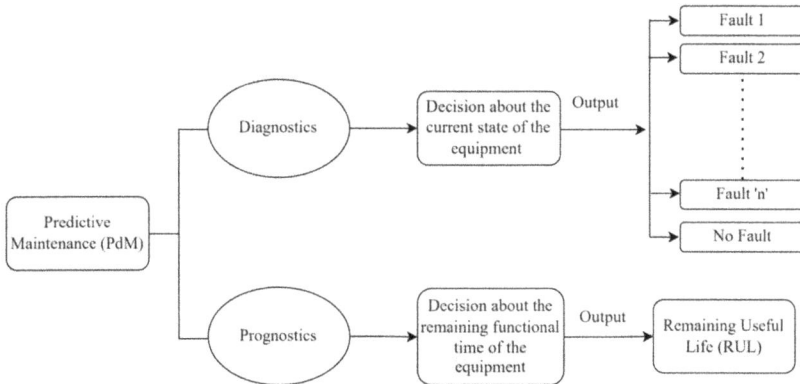

Figure 12.4 The predictive maintenance process, illustration. The output of this process would be either fault detection or estimation of the remaining useful life, based on whether the aim is to carry out diagnostics or prognostics, respectively.

ideas of ML as applicable to PdM. This will help to understand and appreciate the implementation of ML-based PdM in a better way. Section 12.4 will cover several studies where ML-based PdM has been applied for diagnostics and prognostics of engineering systems.

12.2 DIAGNOSTICS AND PROGNOSTICS BASED ON PREDICTIVE MAINTENANCE

Recall that PdM is about deciding when the maintenance is necessary based on the machine's condition through continuous monitoring and analysis of the data. In terms of implementation, we can divide PdM into diagnostics and prognostics (see Figure 12.4). Diagnostics deals with analyzing the current health condition of the equipment to determine whether it is usually working (without Fault) or whether there is some fault (Fault 1, Fault 2, …, Fault "n"). On detection of abnormalities or faults, alarms would be raised. On the other hand, prognostics refers to making predictions of the remaining useful life (RUL) of the equipment using the present and historical data. Here, we mainly convey the information about how long the system will be functional before the maintenance needs to be planned.

12.2.1 Diagnostics

Fault diagnostics through PdM has been widely explored and implemented by researchers and industries across the globe. Let us understand the use of PdM as a diagnostics tool through some examples.

Bearing plays a very critical role in any rotating machinery. Failure of bearings not only affects the machine performance, but may also result in shutting down of the machine, thereby making a costly impact to the organization. Even though a fault in the rotating machine may arise from failure of any of its components, maximum number of such fault scenarios arise from the failure of rolling bearings (41%), as shown in Figure 12.5 (Jha and Swami, 2021). Therefore, diagnostics of bearing faults is an area that seeks a lot of attention.

Figure 12.6 shows a schematic representation of rolling bearing. Each of the four elements, i.e., the rolling element (RE), inner race (IR), outer race (OR), and bearing cage (BC) may develop faults during the operation. VA has been effectively used to diagnose bearing health and detect defects. All the faults have their characteristic frequencies as mentioned in Equations (12.1–12.4) where, D_b is the RE diameter, D_p is the pitch diameter, f_s is the running frequency (Hz), N_b is the number of rolling elements, and "α" is the contact angle between the rolling element and bearing race. If these faults are present in the bearing, they reflect in the vibration frequency spectrum. Fast Fourier transform (FFT) is a commonly used tool that helps to determine the dominant frequencies. Using this information about dominant frequencies, the PdM program identifies the fault by relating these frequencies with the characteristic fault frequencies.

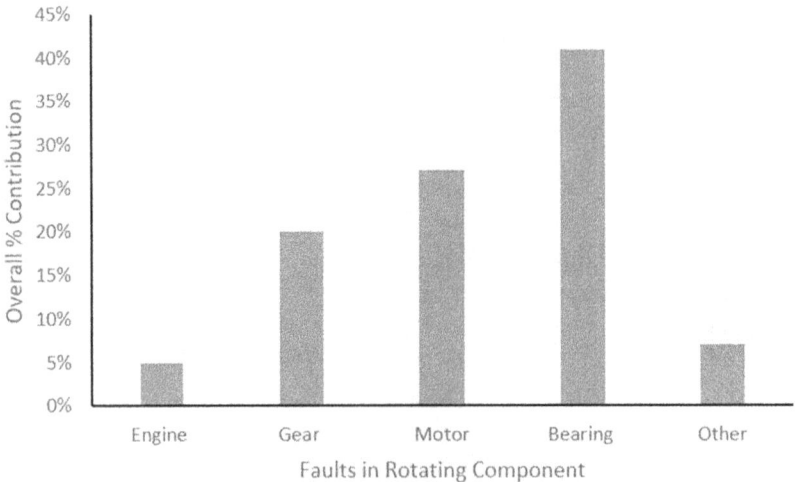

Figure 12.5 Commonly occurring faults in rotating machines and their proportions. The figure is adapted from the data published in (Jha and Swami, 2021). Failure of rolling bearings is the most commonly occurring fault, having a proportion of around 41%.

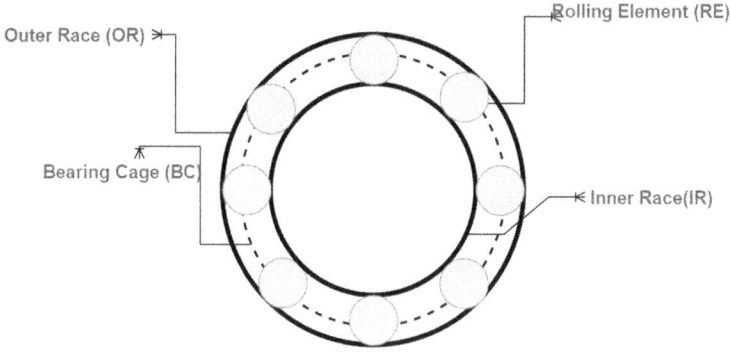

Figure 12.6 Schematic representation of rolling bearing. Failure of rolling bearings can be attributed to fault is any of its four components, namely inner race, outer race, bearing cage, and rolling element.

$$f_{IR} = f_s \frac{N_b}{2} \left(1 + \frac{D_b}{D_p} \cos(\alpha) \right) \tag{12.1}$$

$$f_{OR} = f_s \frac{N_b}{2} \left(1 - \frac{D_b}{D_p} \cos(\alpha) \right) \tag{12.2}$$

$$f_{RE} = f_s \frac{D_p}{2D_b} \left(1 - \frac{D_b^2}{D_p^2} \cos(\alpha)^2 \right) \tag{12.3}$$

$$f_{BC} = f_s \frac{1}{2} \left(1 - \frac{D_b}{D_p} \cos(\alpha) \right) \tag{12.4}$$

Similarly, whenever there is fluid leakage from a pipeline, acoustic waves will be generated and temperature will also change. Acoustic data, temperature data, etc., obtained from appropriate sensors, have been used for detection of leakage in a gas pipeline (Tian et al., 2021).

The idea behind providing these examples is to convey to the readers the main objective of diagnostics in PdM, which is about detecting the presence of any fault in the machine or equipment.

12.2.2 Prognostics

Prognostics typically predict the remaining life span of equipment, asset, machinery, or a component. RUL estimation helps the management plan the resources accordingly depending on where the remaining life expectancy is low (Jiao et al., 2020). We now discuss various methods used for RUL estimation.

Based on the distribution of failure time: Such an approach is feasible when we have information about historical failure times, of the systems similar to the one under study. The probability distribution of this failure time is then used to estimate the RUL. Note that in this approach we do not estimate the RUL in terms of time or number of life cycles remaining, but, in terms of probability of survival beyond a specific time. As shown in Figure 12.7, the probability of failure for the current system after functioning for a time of "t" units is x%, which can also be understood as a probability of survival of (1-x)%.

Based on the degradation profiles: In this approach, we have the complete information, until failure, of the systems similar to the one under study. This information is represented in terms of a degradation profile, based on a health indicator, as demonstrated in Figure 12.8. Here, 1, 2, and 3 indicate the degradation profiles of past systems with respective failure times t1, t2, and t3, while 4 indicates the degradation profile of the present system.

For demonstration purpose, we have chosen only a few degradation profiles, but in reality there will be many more. RUL in this case is estimated based upon that degradation profile which matches closely with the degradation profile of present system. In Figure 12.4, since profile 4 matches closely with 2, RUL is the difference between the failure time of a similar

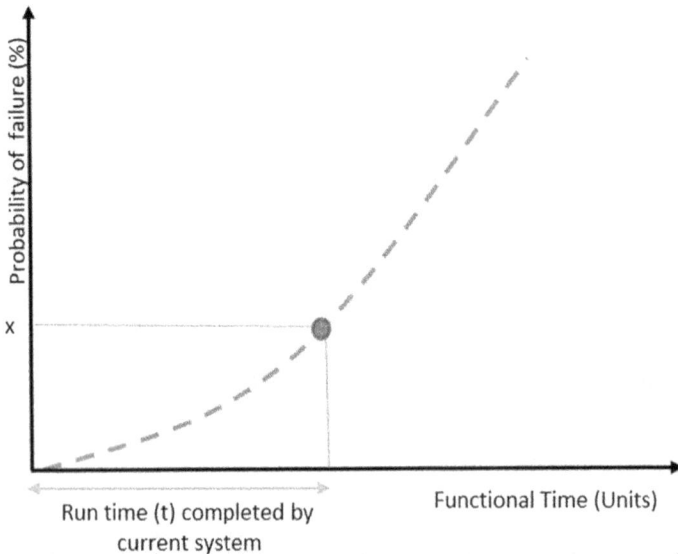

Figure 12.7 Probability distribution of failure time. This is used to estimate the probability of survival/working of an equipment or a component beyond a particular time.

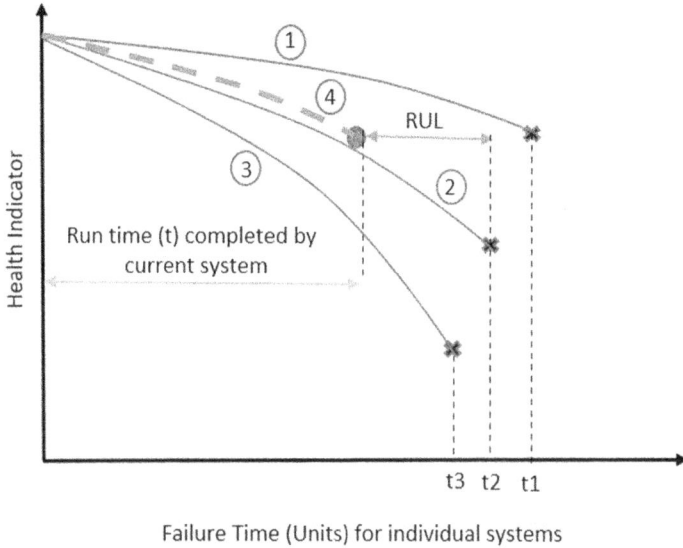

Figure 12.8 RUL estimation based on degradation profiles. The degradation profile of current system (4) matches closely to that of system (2). Therefore, the RUL estimate can be made as the difference between t2 and t.

system 2 (t2) and run time (t) completed by the current system. This approach is implemented by Khelif et al. (2015), where discharge characteristics of lithium-ion batteries have been used to estimate the RUL of batteries. This study was based on dataset provided by NASA. The dataset was built by accelerated aging experimentation, which was carried out until the test batteries failed. The entire health information until the failure time was recorded and transformed into degradation profiles, as we discussed.

Based on threshold Levels: This method becomes significant as practically it may not always be feasible to have information about functional lifetime, or also the failure times may not be available. The principle behind this method is based on knowledge about a certain level beyond which failure would be inevitable. Such a level is called a threshold level. RUL estimates, in this case, would contain information about how much time is available until the threshold level is crossed. Figure 12.9 shows one such example. RUL estimation has been studied based on threshold levels for oil and gas transportation systems by Cai et al., 2021. The authors considered actual working conditions and expert opinions to decide their threshold level.

Diagnostics and prognostics, therefore, are two branches of a PdM strategy, each having a distinct objective. Choice between diagnostic and prognostics approaches, however, will be depending upon the problem to be solved.

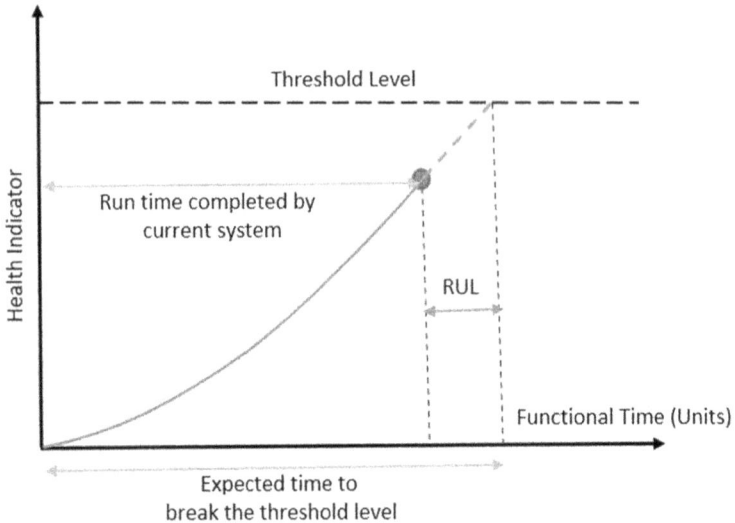

Figure 12.9 Threshold-based approach for RUL Estimation.

12.3 MACHINE LEARNING FOR PREDICTIVE MAINTENANCE

As we realize that PdM is primarily a data-oriented technique, collection and processing of the data play a pivotal role in the final output of diagnostic or prognostic approaches (see Section 12.2). Data collection refers to recording different measurements from the machine or equipment under a variety of operating conditions as may be required from case to case. With the advances in sensor technologies, data collection is now more feasible than earlier. Different types of measurements can be made using respective sensors. Few examples of measurements and the associated sensors would be:

- Measuring vibrations using piezoelectric accelerometers
- Sound measurement using microphones
- Temperature measurement using resistance temperature sensor, thermocouple or infrared thermography
- Measurement of motor current using current transformers
- Oil quality measurement using particle counter
- Flow pressure measurement using pressure transducers

Once the data is collected, the next task is to be able to use this data to determine the output of diagnostic and prognostic approaches in PdM (see Section 12.2). In recent years, people have looked at ML techniques as an alternate way to make decisions based on the data. Figure 12.10, for

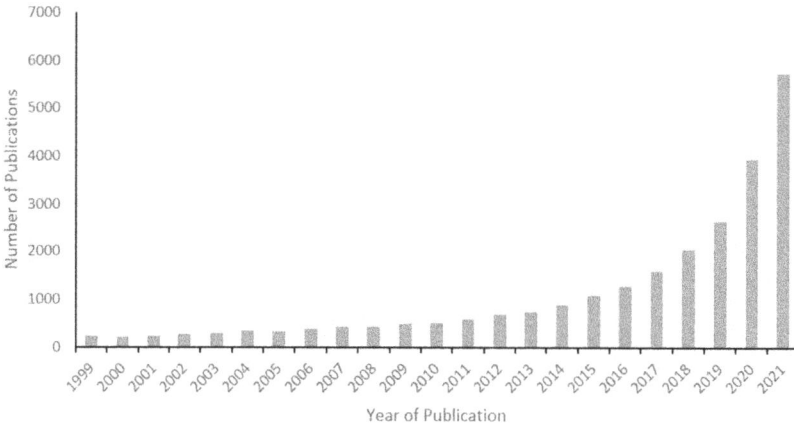

Figure 12.10 Number of publications in *Machine learning-based predicting maintenance* in Science Direct Database. The search was based on keywords: "Predictive Maintenance" and "Machine Learning."

example, shows that there is a steep rise in the number of articles published in the field of PdM using ML methods, hereafter referred to as ML-based PdM.

We now give a brief overview of ML methods. ML methods are typically classified into supervised learning, unsupervised learning, and reinforcement learning at a broad level (Figure 12.11). We consider reinforcement learning out of scope in our discussion.

Supervised learning methods learn a function that describes the relationship between training data samples' features and outcome value. A training data sample (x, y) consists of features $x_1, x_2, \dots x_D$ and an outcome variable y. For example, in a food processing company, a refrigeration unit is required for cold storage of the products, which otherwise may get spoiled. We may use recorded measurements such as temperature and sound pressure levels to predict whether the refrigeration unit is normal or faulty. In this case, the recorded measurements are features, and healthy/faulty conditions are the outcomes to be predicted. Also, each recorded data point is called a sample, and we assume that we have "n" such samples for training the ML model. Note that we know the outcome value or the class label for each sample. The end goal of this learning process is that once we learn the function $f: x \rightarrow y$, we can predict the outcome value for a newly encountered data sample x, with an unknown class label.

Supervised learning methods are used to solve two types of problems: classification and regression. In a classification problem, the goal is to predict a class y from several possible classes, $y \in \{0, 1, \dots, k\}$. For example, we would want to predict the correct condition of a rotating machine from possible options such as normal, bearing fault, unbalance, and shaft bent. On the other hand, in a regression problem, the aim is to estimate the future

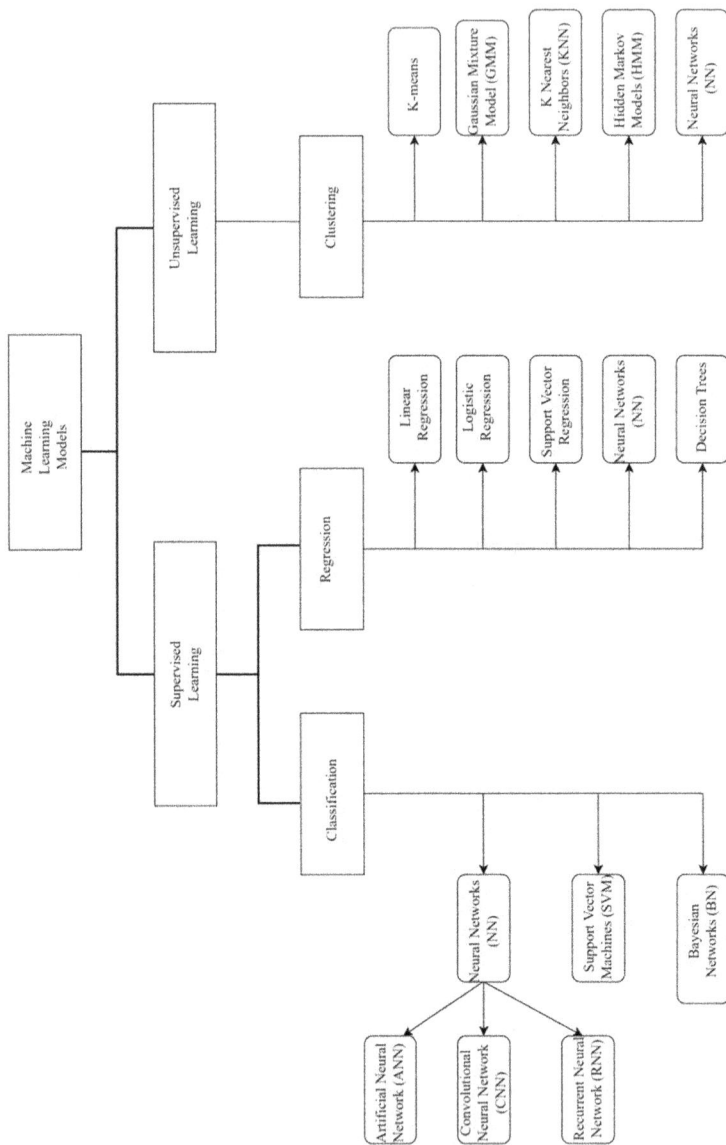

Figure 12.11 Categorization of various machine learning methods—not a comprehensive list, but gives a bird's-eye view of methods.

value of an output variable, i.e., $y \in R$ (real valued numbers). For example, consider the case where we want to determine the amount of wear in the cutting tool. Here, the main objective would be to develop a regression model which can forecast the amount of wear in the cutting tool using some data measurements. We choose a classification or regression-based model based on the diagnostics and prognostics requirement.

In an unsupervised learning setting, we deal with x but no outcome variable y. That means learning algorithms learn a structure in the data without having any prior information about output classes (Cinar et al., 2020). For example, such algorithms learn patterns in the data and form clusters of training data samples with similar characteristics. The new test samples are assigned to a particular cluster based on this learning. Let us consider an example of wind turbine anomaly detection. We assume that the recorded historical measurements do not have output labels. Thus, supervised learning models will not be feasible in this case. However, an unsupervised learning model will learn the patterns in this unlabeled data and segregate the various samples into different clusters. Each of these clusters ideally represents different wind turbine operational conditions. For example, the unsupervised learning algorithm may find two patterns in the data, belonging to the faulty and regular operation of the wind turbine and, therefore, form two clusters.

Supervised learning methods are popular for building ML-based PdM pipelines, some of which we will see in Section 12.4. However, interested readers can refer to Carasco et al. (2021) and Kolokas et al. (2020) for motivating the applications of unsupervised learning methods in PdM. For a detailed discussion on the working of various ML algorithms, see, for example, Bishop and Nasrabadi (2006) and Hart et al. (2000).

12.3.1 Machine learning-based predictive maintenance pipeline

In this section, we discuss an ML-based PdM pipeline. For ease of discussion, we divide the PdM pipeline into four units: A, B, C, and D (see Figure 12.12 for more details). Target system represents particular equipment or a component or an entire system where we plan to implement PdM. For example, in the heating, ventilation, and air conditioning (HVAC) system, the entire HVAC unit becomes our target system for applying PdM. If we want to implement PdM for an automobile engine, then the engine becomes our target system. As shown in Unit A, we first collect the data from the target system. We need historical data to train and test an ML model. In addition, the present running condition data, which we refer to as new data, will be used to make predictions about the present state of the system or equipment.

In Unit B, we preprocess the data collected, historical as well as current data. This is done to remove any discrepancies in the recorded data, thereby

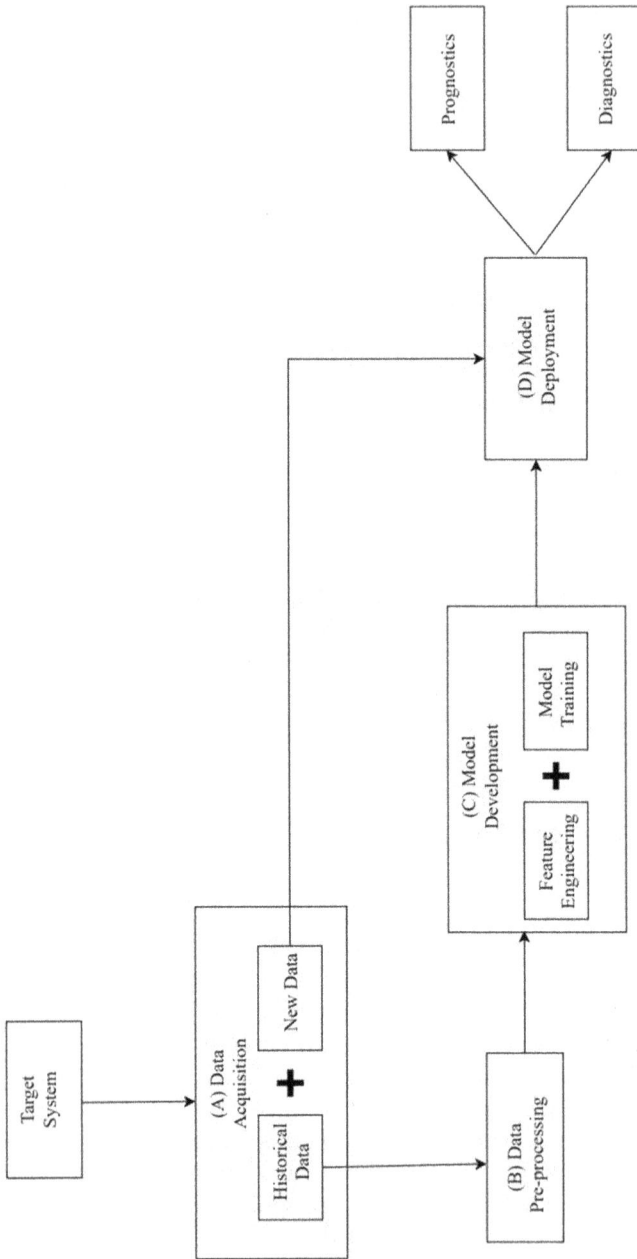

Figure 12.12 A machine learning-based PdM pipeline. Historical data from the target system is used to train the ML model. The present or new data is fed to the trained model to make predictions as per diagnostics and prognostics tasks.

providing clean data. One such method to clean the data is to remove any outlier elements by using 6σ criterion (Equation 12.5). Here, for a feature x, only the data which lies within the $(\mu + 3\sigma)$ and $(\mu - 3\sigma)$ limits is retained, while the rest is discarded. Here, μ is the mean and σ is the standard deviation for any feature x. Assuming normality of feature x, we can see that 99.73% of the data will lie within the said limits.

$$P(\mu - 3\sigma < x < \mu + 3\sigma) = 0.9973 \tag{12.5}$$

We also scale or normalize the data to remove the influence of several features with a larger magnitude over those with smaller ones. Two popular data scaling methods are: (1) standard scaling (Equation 12.6), which converts each feature to a standard normal variate, and (2) min-max scaling (Equation 12.7), which converts each feature between 0 to 1. Here, x' represents the scaled feature, x represents the original feature, and μ represents the mean of the original feature.

$$x' = \frac{x - \mu}{\sigma} \tag{12.6}$$

$$x' = \frac{x - min(x)}{max(x) - min(x)} \tag{12.7}$$

Unit C deals with building the ML model and has two parts. First, we select those features which will make a significant contribution in yielding better prediction from the model. Note that we may have a large number of features recorded during the data acquisition phase, resulting in a high dimensional data. High dimensionality, even though it sounds exciting, often acts as an obstacle for achieving high model performance, and is signified in what we call as *curse of dimensionality* (Berisha et al., 2021). Therefore, picking out important features is an essential step, thereby reducing dimensionality and is referred to as feature selection. Feature extraction also helps to reduce the number of features. However, unlike feature selection, it does so by explaining the data in terms of new statistical features while still retaining the original information.

Both of them feature selection and feature extraction together form the first part of Unit C, labeled as feature engineering. The preprocessed data is then split into two parts: (a) training and validation set and (b) test set, both of which contain only those features selected/extracted through feature engineering. Using the training and validation set, ML model is trained to learn the function f: $x \rightarrow y$, while the test data is used to evaluate the model's performance. Next section will provide explanations on ways of evaluating performance of an ML model.

Once the model achieves an acceptable level of accuracy, we deploy the trained model to make actual predictions, which is represented by Unit D.

12.3.2 Evaluating machine learning models

Before deploying an ML model to make actual predictions, we need to evaluate the model's performance. Different evaluation criterions are used for classification and regression problems.

For regression models, the performance is typically quantified via metrics such as mean absolute error (MAE), mean squared error (MSE), root mean squared error (RMSE), and R-squared (R^2), as described by Equations (12.8–12.11). We denote the actual or "predicted" value and the "true" value of the output variable for the ith sample by y_i and $\hat{y}_i$ respectively. Here, n denotes the total number of samples, and $\bar{y}$ indicates the mean of all values. The aim is to determine the closeness of model prediction with the true value, and this is evaluated based on the test dataset.

$$MAE = \frac{1}{n}\sum_{i=1}^{n}|y_i - \hat{y}_i| \qquad (12.8)$$

$$MSE = \frac{1}{n}\sum_{i=1}^{n}(y_i - \hat{y}_i)^2 \qquad (12.9)$$

$$RMSE = \sqrt{\frac{1}{n}\sum_{i=1}^{n}(y_i - \hat{y}_i)^2} \qquad (12.10)$$

$$R^2 = 1 - \frac{\frac{1}{n}\sum_{i=1}^{n}(y_i - \hat{y}_i)^2}{\frac{1}{n}\sum_{i=1}^{n}(y_i - \bar{y})^2} \qquad (12.11)$$

On the other hand, classification models are evaluated based on correct and incorrect classification performance. For example, consider the safety valve mounted on the boiler to release excess fluid pressure. To build a classifier, let us say that failure of this safety valve is a positive class, while proper functioning is a negative class. We first define four count-statistics that help define popular evaluation methods. True positive (TP) instance means the classifier correctly identifies the class of the test sample as the positive class. False positive (FP) instance means the classifier incorrectly classifies the test sample as the positive class. True negative (TN) instance means the classifier correctly identifies the class of the test sample as the negative class. False negative (FN) instance means the classifier incorrectly classifies the test sample as the negative class.

The total number of *misclassifications* in a validation or test set is computed by counting all such FP and FN instances. Four popular evaluation scores (Hossin and Sulaiman, 2015) based on the above count-statistics are accuracy, precision, recall, and F1-score, which are described in Equations (12.12–12.15).

$$Accuracy = \frac{TP + TN}{TP + TN + FP + FN} \tag{12.12}$$

$$Recall = \frac{TP}{TP + FN} \tag{12.13}$$

$$Precision = \frac{TP}{TP + FP} \tag{12.14}$$

$$F1 - Score = 2 * \frac{Precision * Recall}{Precision + Recall} \tag{12.15}$$

12.4 MACHINE LEARNING-BASED PREDICTIVE MAINTENANCE IN ENGINEERING SYSTEMS

ML has been a promising tool for developing PdM strategies for different industrial sectors in recent years. As mentioned in Section 12.3, people have explored the scope of ML based PdM approach for diagnosis and prevention of failures. In this section, we discuss case studies in different engineering systems that used ML-based PdM with the pursuit of yielding superior results.

12.4.1 Machining systems

Machining systems, in this chapter, refer to the machines or equipments involved in various machining operations such as turning, milling, grinding, etc. All such systems are an assembly of components like cutting tools, electric motors, bearings, spindles, and many more. Failure of any of these constituent components leads to the failure of the machining system and therefore invariably affects the production requirements. We will look at the application of ML-based PdM of the components of machining systems for detection and/or prediction of failures. As the list is large, we focus only on the following components in this discussion:

- Cutting tool
- Bearing

12.4.1.1 Cutting tools

Cutting tools typically remove material from a workpiece to give it the desired geometry. While the workpiece material is plastically deformed, the cutting tool experiences wear, which will grow over time. As the cutting tool wears out, it will affect the product quality, and hence the condition of the cutting tool must be appropriately monitored. The amount of wear on the different faces of the cutting tool determines its state. ML-based PdM approach can be applied to diagnose the condition of the cutting tool used in machining processes. One such strategy is using the length of flank face wear as the criteria (Lee et al., 2019). For example, on a cast-iron workpiece (480 mm × 178 mm × 51 mm), using the length of the flank face wear as the criteria, the condition of the cutting tool can be classified in three stages: normal, warning, and failure. This makes it a multi-classification problem.

To solve this multi-classification problem, (Lee et al., 2019) demonstrated the use of support vector machine (SVM) classifier. SVM is a supervised learning method that separates data samples belonging to different classes using a hyperplane. It takes input at the D-dimensional feature vector $x_1, x_2, ..., x_D \in R^{\wedge}D$ with the associated output class labels $y_1, y_2, ..., y_k$. The minimum distance between the samples in different classes is called *margin*. The objective is to identify the hyperplane that maximizes the minimum distance between the samples in different classes (Wang, 2005). It is also important to understand what type of data is required for such problems, and the same is presented in Table 12.1. Let us understand the effectiveness of such an approach in detecting the condition of the cutting tool through a confusion matrix (see Figure 12.13). Confusion matrix represents the model performance on the test dataset in a 2D matrix format. On one axis actual class is mentioned, while on the other predicted class is included. The

Table 12.1 Summary of the test conditions used by Lee et al. (2019)

Parameter	Specification
Speed	826 RPM
Depth of Cut (2 Different Values)	0.75, 1.5 mm
Feed (2 Different Values)	0.25, 0.5 mm/rev
Signals Monitored	Motor Current, Vibration, Acoustics

Note: These recorded measurements formed the historical data as described in Section 12.2

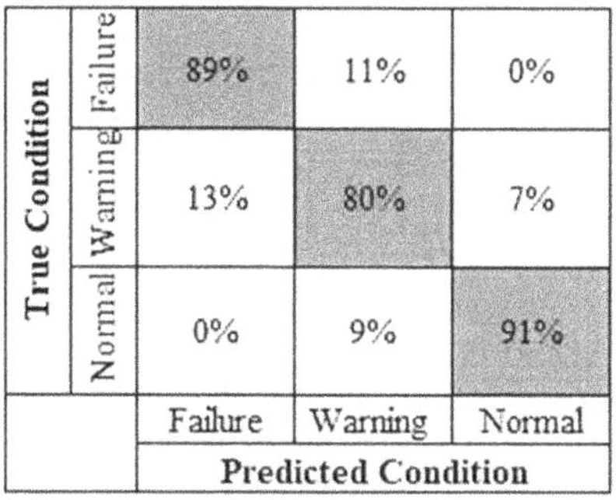

	89%	11%	0%
	13%	80%	7%
	0%	9%	91%
	Failure	Warning	Normal

Predicted Condition

Figure 12.13 Confusion matrix depicting classification results of support vector machine (SVM) classifier model for cutting tool conditions (Lee et al., 2019).

diagonal elements represent the matching of predicted and actual class, representing correct classifications, while the off-diagonal elements represent the misclassification or "confusion." In this case, we can observe that SVM classifier yields a decent classification result with a maximum 13% confusion that exists between warning and failure mode.

12.4.1.2 Bearings

Bearings form a vital and integral part of the rotating machinery. They perform several functions such as reducing friction, ensuring smooth rotation of shaft, and providing support reactions. Like all the components, bearings too are likely to develop different faults, which we have discussed in Section 12.2. Failure of bearings alone can be responsible for the shutdown of a machine or an equipment. We will now look at how ML-based PdM can be applied for fault detection of bearings.

Detection of bearing health conditions can be done using vibration data in conjunction with deep learning (DL) techniques. Before further discussing bearing fault detection, let us understand what DL is. DL is a subset of ML that involves a multi-layered artificial neural network (ANN). The working of ANN is inspired by how the nervous system of the human brain functions. A basic ANN consists of input, output, and hidden layers. Each layer consists of neurons, which resemble brain cells. A mathematical representation of the biological neuron is termed as "perceptron," and thus, multi-layered neural network is also referred to as multi-layer perceptron (MLP) model. The hidden layer develops interconnections between the neurons to establish a relation between the input feature and output class. An ANN with more than one hidden layer is termed a "deep neural network (DNN)." We will now look at two different DNNs and their applications in bearing fault detection. First is a convolutional neural network (CNN), a DL method

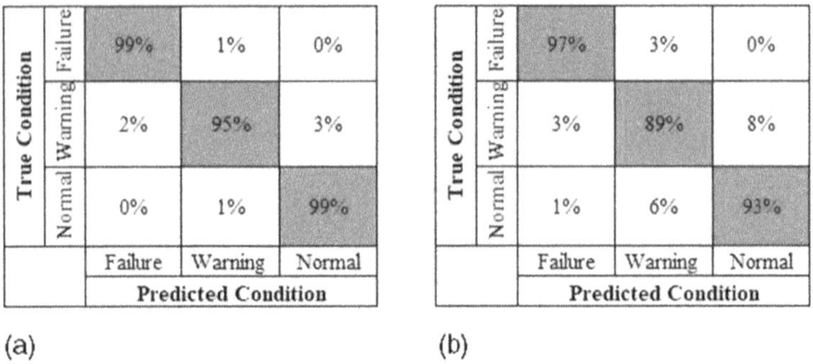

True Condition		Predicted Condition	
Failure	99%	1%	0%
Warning	2%	95%	3%
Normal	0%	1%	99%
	Failure	Warning	Normal

(a)

True Condition		Predicted Condition	
Failure	97%	3%	0%
Warning	3%	89%	8%
Normal	1%	6%	93%
	Failure	Warning	Normal

(b)

Figure 12.14 Classification of bearing faults using two deep learning approaches: (a) convolutional neural networks (CNNs) and (b) recurrent neural networks (RNNs) (Lee et al., 2019).

that primarily uses images as input. CNN converts a raw image into a feature vector through different operations (see, for example, Albawi et al., 2017). A layered neural network architecture then processes the feature vector to classify the image into a particular class. Recurrent neural network (RNN) is another DL algorithm that has gained popularity owing to its capacity in processing sequential data. This property makes it an essential model for solving problems where the data is time instance based. Fundamentally, RNN is a neural network consisting of multiple layers where the information flows in a cyclic loop. It has a memory storage property, by which it uses the information gained from previous input while processing the current input. This memory storage property allows it to process sequential data rather than demanding entire data at once, as regular neural networks do. A comprehensive description of the working of RNN can be found in Sherstinsky (2020).

An ML-based PdM approach using CNN and RNN for bearing fault detection has been demonstrated by (Lee et al., 2019), where remaining functional life is used as deciding criteria. Based on the value of remaining functional life, we will be able to classify the bearing condition as normal, warning, or failure. For this, however, we will need a dataset containing measurements until the bearing failure. The vibration data recorded through accelerometers at a particular sampling rate can be used as an input for CNN and RNN models. For example, in (Lee et al., 2019), the sampling rate was chosen as 20 kHz, and results obtained from their experimentation can be seen in Figure 12.14.

Another type of data that can be used for detecting bearing faults using ML is temperature images. The process of developing thermal images from a machine or an equipment is called *thermography*. This help to detect not only bearing faults but faults in other components of a machine as well. For example, Yongbo et al. (2020) used thermal images as input for classification of nine different faults including bearing faults as given below:

- Unbalance
- Fault in the inner race of bearing
- Fault in the outer race of bearing
- Fault in the ball of the bearing
- Presence of combination of bearing faults
- Presence of unbalance along with fault in the bearing inner race
- Presence of unbalance along with fault in the bearing outer race
- Presence of unbalance along with fault in the bearing ball
- Presence of unbalance along with combination of bearing faults

An overview of this two stage ML model can be seen in Figure 12.15. Since, while using thermography, our input data will be in the form of temperature images, CNN can be used to extract relevant features from these images. A multi-layer neural network can then use these features for classification of faults.

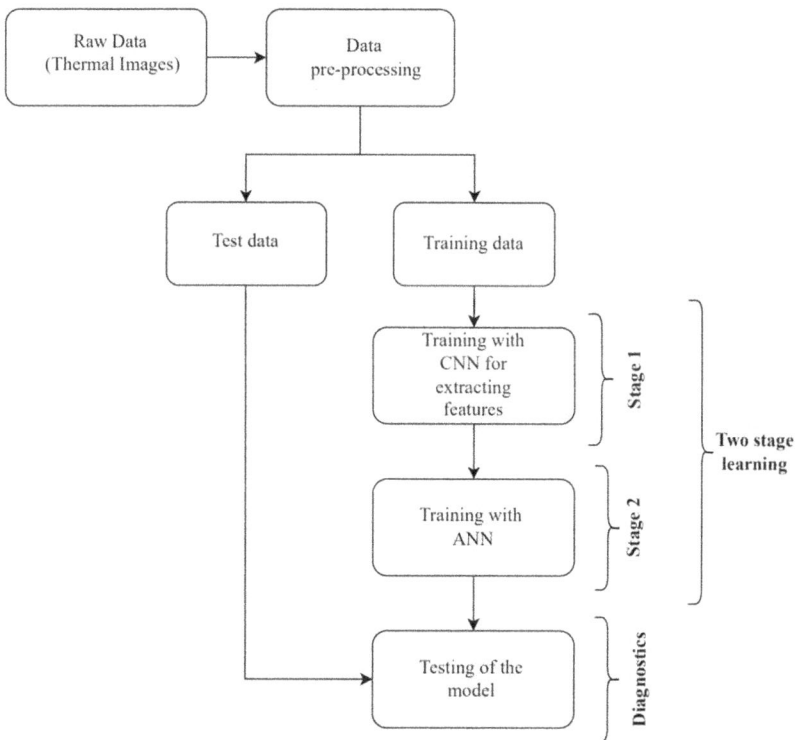

Figure 12.15 Two-stage learning model with convolutional neural network and artificial neural network for bearing fault detection, using thermal images (adapted from Yongbo et al. (2020)).

Thus, we can now say that application of ML for bearing fault detection or prediction of bearing condition is a promising technique.

12.4.2 Automotive systems

In this chapter, automotive systems refer to systems involved in power transmission and motion control of a vehicle. Since such systems would form the heart of entire vehicle, their reliability should be high. Avoiding any kind of failure is, therefore, a primary objective to ensure that any fatal accident does not occur. The fusion of ML and PdM can contribute significantly toward early detection of any failures of such system, thereby providing time for repair and replacement. We will look at two of the major systems in a vehicle:

- Brake system
- Gearbox system

12.4.2.1 Brake system

Brake system forms a critical part of an automotive product. As we all know, its function is to slow down or stop a moving vehicle. Obviously, we would want that braking system on any vehicle should function properly, and therefore proper maintenance of such a system is very crucial. Having said that, braking system is a complex assembly of multiple components such as brake pad, liners, brake pedal, cylinders, disc, etc. ML-based PdM can be one solution to detect any faults in the braking system. For example, fault detection in the hydraulic braking system of a vehicle can be done using ML-based PdM (Pranesh et al., 2021).

Since, failure of more than one component in the braking system need to be identified, it would result into a multi-classification problem. Therefore, we can find the following classes being considered for fault classification (Pranesh et al., 2021):

- GOOD: Brake components in good condition
- AIR: Presence of air in the brake fluid
- BOS: Spillage of brake oil
- EIO: Wear of brake pad in an even manner on inner and outer side
- EI: Wear of brake pad in an even manner on inner side
- UI: Wear of brake pad in an uneven manner on inner side
- UE: Wear of brake pad in an uneven manner on inner and outer side

What types of data and ML model are suitable for such problems? One answer could be vibration measurement as the data, along with logistic regression (LR) as the ML model. Even though the name suggests regression,

LR is a popular supervised learning classification model. It performs well in binary as well as multi-classification problems. As we discussed in Section 12.3, this supervised learning model learns the function f: $x \to y$ and assigns a probability for each possible output class. Based on the highest probability, a sample will be fit to a particular class. For more detailed description, we recommend reading Healy (2006).

Let us now look at confusion matrix of the results obtained by (Pranesh et al., 2021) using vibration data (see Figure 12.16) and LR classifier. In this matrix, we have a total of 515 samples (sum of all the elements) from which 451 instances have been identified correctly (sum of diagonal elements). Thus, the overall accuracy of correct classifications is 87.57% (ratio of correctly classified instances to the total number of samples). Individual class-wise statistics can be noted from Table 12.2.

Another issue regarding the braking system is of brake squeal (noise from the braking system). During the engagement of the brake disc and the pad, friction may generate, giving rise to brake squeal. Reducing it, to meet the acoustic norms, is a challenge for the brake manufacturers (Wu et al., 2021).

		GOOD	AIR	BOS	EIO	EI	UI	UE
Predicted Label	GOOD	67	0	0	0	0	6	2
	AIR	0	67	0	0	0	6	2
	BOS	0	0	68	6	0	1	0
	EIO	0	0	6	69	0	0	0
	EI	16	0	0	0	59	0	0
	UI	0	10	3	3	0	61	1
	UE	0	5	0	0	0	3	70
		GOOD	AIR	BOS	EIO	EI	UI	UE
				Actual Label				

Figure 12.16 Confusion matrix depicting classification performance of logistic regression model on predicting the brake fault conditions.

Table 12.2 Class-wise statistics for brake fault detection using logistic regression

Class	Precision	Recall	F1-score
Good	0.775	0.733	0.753
AIR	0.817	0.893	0.854
BOS	0.883	0.907	0.895
EIO	0.92	0.92	0.92
EI	0.747	0.787	0.766
UI	0.897	0.813	0.853
UE	0.959	0.933	0.946

Source: The results are adapted from (Pranesh et al., 2021)

ML-based brake squeal prediction is studied by Stender et al. (2021) employing DL-based algorithms.

It is important to note that the perception of sound is frequency dependent and, hence, identifying squeal correctly is a challenge. We can formulate this into a multi-classification problem considering different types of sounds as, for example:

- Squeal: tonal sound
- Click: impulsive sound
- Wire brush: short time sounds at different frequencies
- Artifacts: broadband frequency noise

CNN can be a useful tool here, which can be fed with sound pressure level (SPL) maps obtained from the microphone recordings as input. The classification, in such cases, is performed based on machine vision principles. Looking at results mentioned by Stender et al. (2021), the highest mean average precision value of 80% is recorded with this approach. If, instead of SPL maps, the prediction has to be made using time series data from the microphone readings, RNN can be an effective tool. Stender et al. (2021) also explored this approach and have quantified classifier performance using Mathews correlation coefficient (MCC), with highest MCC being around $0.78+/- 0.02$. MCC is more advantageous than the F1 score (Chicco and Jurman, 2020) that we discussed in Section 12.3. MCC ranges between $[-1, +1]$, with $+1$ and -1 indicating perfect classification and misclassification.

12.4.2.2 Gear system

Gears are widely used as power transmitting members. Due to variation in operational conditions, gears are subjected to various faults. Following are some commonly occurring faults in the gear drive system:

- Error in the gear tooth profile
- Cracking of the gear surface and gear tooth
- Error in alignment

Being a critical member in the power transmission unit, failure of gear system would halt the motion of entire vehicle. Therefore, it is essential to detect these faults to prevent any significant failure in the power train.

Most of the times decisions about faults present in the gears are based on perception of the sound emitted by gears. Psycho-acoustic features, for example roughness and loudness, describe the perception of sound by human ears. Kane and Andhare (2020) demonstrate the use of psycho-acoustic features for gear fault detection based on ANN (Table 12.3).

It is also important to realize that prediction of future failure is also as equally significant in detecting faults. Since such prediction involves time-based

Table 12.3 Performance of different features in predicting the gear faults using artificial neural networks (ANN)

Features	Training accuracy (%)	Testing accuracy (%)
Acoustic	68.58	64.56
Vibration	94.67	89.06
Psycho-acoustic	97.99	95.93

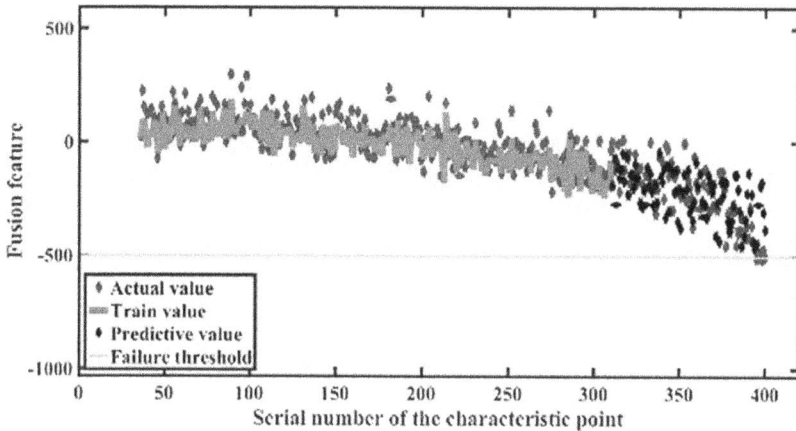

Figure 12.17 Remaining useful life prediction of gear using recurrent neural network. The predicted degradation curves are found to be closely matching with the actual degradation curves, indicating the effectiveness of this method (Xiang et al., 2020).

forecasting, RNN can be a suitable choice of model. One case study in this regard has been presented by Xiang et al. (2020). An indicator of gear health can be established from the recorded measurements and an appropriate threshold level (based on expert knowledge or past failure data) be assigned to this health indicator. RNN will predict the degradation profile based on health indicator values. RUL can then be estimated by comparing these health indicator values with the threshold level set (see Figure 12.17). We have discussed this approach of RUL estimation in Section 12.2.

12.4.3 Production systems

In any manufacturing industry, the focus is on ensuring continuous functioning of the production line. Often termed as *transformation process*, a production line will transform the raw material into final, useful products. It consists of several machines operating in co-ordination and executing specific tasks to manufacture the product. Faults in any of the constituent machines of this production line will have an adverse effect on the production, quantitative (number of units produced) as well as qualitative (number of defective units). Of course, this is a direct impact on operational costs for any organization, and, therefore, maintaining a production line in proper working condition is something that cannot be ignored.

In order to apply ML-based PdM in such manufacturing units, we will need to collect data from all the machines that form the production line. For this, different sensors, some of which we mentioned in Section 12.3, need to be installed on each machine. For example, Ayvaz and Alpay (2021) demonstrated the implementation of ML-based PdM on the production line of a baby diaper manufacturing plant. The authors aimed at detecting RUL between two shutdowns of the production line occurring due to two faults. The two faults are related to failure of the raw material feeding machine, classifying them as Weight Low 1 and Weight Low 2. IoT sensors, mounted on the machines, recorded measurements about speed, motion, temperature, electric current, weight, vacuum, and air pressure.

In such cases, we also want to inform the readers about two challenges that need to be overcome.

The first one arises from having a large number of different recorded measurements, i.e., different features. We call such type of data as high-dimensional data. It is desirable to covert the data in low-dimensional configuration to improve the model performance. Principal component analysis (PCA) is one technique that performs this dimensionality reduction task. PCA identifies the underlying correlations between a large number of features. These correlations are called principal components. All the original features are than transformed into new form using these principal components, resulting in lower dimensionality. It is important to note that even in low dimensional form, the information content of the original data is retained. In the case of diaper production unit (Ayvaz and Alpay, 2021), application of PCA yielded a new 17-dimensional data, from the original set of 101 features. These 17 new dimensions explain 95% of the variation in the original dataset.

The second challenge arises from having imbalanced data. Imbalanced data means that number of training samples in one class is very less (minority class) compared to other classes. This affects learning of the model and makes it more biased toward the class containing large number of samples (majority class). Many methods have been developed by researchers to address this issue (Rout et al., 2018). Ensemble learning is one method where multiple data sub-sets are formed by resampling the original data. While doing this, the idea is to use all samples of minority class while picking certain samples from majority class, to create a balanced sub-set. The final decision is obtained after considering the predictions obtained on each sub-set. Two of such ensemble algorithms, bagging and boosting (Buhlmann, 2012), have been applied to balance the dataset obtained from the production line (Ayvaz and Alpay, 2021). Moreover, it can be a good strategy to evaluate the performance of various models as this performance will vary

Table 12.4 Evaluation of different algorithms based on binary fault classification in a production line

Algorithm	R^2	MAE	MAPE	RMSE
Random forest	0.982	51.97	3.27	147.19
XGBoost	0.979	82.09	5.16	5.16
Gradient boosting	0.776	394.52	24.79	523.91
MLP	0.675	466.32	29.3	632.69
SVM	0.347	682.43	42.88	896.07
AdaBoost	0.338	752.95	47.32	902.85

Source: (Ayvaz and Alpay, 2021)

application-wise. To support this belief, look at the results populated in Table 12.4 using the evaluation criteria discussed in Section 12.3. Here, random forest (RF), XGBoost, gradient boosting and AdaBoost, all of them, are ensemble-based classifiers. Decision tree (DT), which we will discuss in the next section, is another such example. They all apply the principle of ensemble learning, discussed earlier, for classification and regression problems. Please note that these results are claimed by Ayvaz and Alpay (2021), where they mainly solved a regression problem of estimating the RUL, and hence the choice of these evaluation metrics.

12.4.4 Thermal systems

Here, systems that mainly deal with processing of heat energy are referred to as thermal systems. Power generation is one area where thermal systems play a pivotal role, for example boilers, which generate steam. This steam is then used to drive a steam turbine, thereafter converting it into electricity. Another application where we find thermal systems at the core is in the air-conditioning and refrigeration sector. Here, the need for such systems arises from the requirement to maintain proper ambient conditions, for providing comfort, or, for storage of products like food items, medical drugs, etc. A failure of such systems may lead to potential losses to an organization, and, hence, detection of these failures is a necessity. We will now discuss the application of ML-based PdM for two of such thermal systems:

- Boilers
- Chillers

12.4.4.1 *Boilers*

Boilers are enclosed pressure vessels, mainly designed to produce steam, by supplying heat energy to water or any other fluid. It is made up of several components, for example burner, combustion chamber, heat ex-changer, etc.

In such large-scale thermal systems, physical inspection of different components is impossible. Thus, fault detection or diagnostics tasks in these systems
become complicated. Also, one challenge faced in developing an ML-based
PdM tool is the lack of good-quality data describing different faults. To tackle
this issue, data can be simulated from the physics-based model. Shohet et al.
(2020) used simulated data for fault detection in a non-condensing boiler.
For this, the physics-based model is prepared in MATLAB®/SimScape,
using the specifications derived from the Viessmann Vitorond 200 Series
and Raypak Series boilers. While developing an effective fault identification
model for boilers, we have to consider faults in different components of the
boiler. Shohet et al. (2020), for example, considered three faults: excess air
(X), fouling (F), and scaling (S). More details are presented in Table 12.5.
Like we mentioned previously, that, it is a good practice to evaluate the performance of different models to solve the classification problem, Table 12.6
shows the performance comparison of k-nearest neighbor (k-NN), DT, RF,
and SVM. We have earlier discussed in brief about DT, RF, and SVM models.
k-NN model is another supervised learning model, fit for both classification
and regression problems. Here, nearest neighbors for a sample are the data
points closest to it, while k is the number of such neighbors under consideration. Like any other supervised learning model, k-NN first classifies the

Table 12.5 Summary of simulated faults and recorded variables
to develop PdM model for boiler fault diagnostics

Fault Label	Component	Variable
X	Boiler combustion chamber	Airflow rate
F	Gas water heat ex-changer	Fouling factor
S	Gas water heat ex-changer	Scaling factor

Source: (Shohet et al., 2020)

Table 12.6 Performance of various classifiers on Raypak and
Viessmann boiler series models diagnostics

Algorithm	Raypak (%)	Viessmann (%)
k-NN	87.9	92.4
Decision trees	95.4	97.2
Random forests	86.3	92.8
Support vector classifier	82.3	69.6

Source: (Shohet et al., 2020). The numbers denote classification accuracy for
each model

training data samples into various classes. Later, for a test data sample, it determines the classes of its k-NNs. Based on the class to which these majority of k-NNs belong, the test sample is assigned that particular class. For more reading on k-NN, see, for example, Peterson (2009).

12.4.4.2 Chillers

Chillers, a part of HVAC systems, carry heat away from the conditioned space using refrigerant. The chilled refrigerant coming from the evaporator is pumped around the space to be conditioned. From there, it collects the heat and sends the same to the condenser. And finally, the condenser sends this heat into the atmosphere. This brief description helps us to get an idea about the complexity of chiller system. Many faults can be associated with chillers. See Table 12.7 adapted from Han et al. (2021). Here, we can realize the potential of deep neural networks for chiller fault diagnosis. Deep neural networks with simulated annealing (SA-DNN) method proposed by Han et al. (2021) is presented in Figure 12.18. This approach yields an accuracy of 99.3% in detecting seven commonly occurring faults in chillers mentioned in Table 12.7. We would like to inform the readers that the data set used in this study is not simulated from physics-based model; rather, it is obtained through experimentation by American Society of Heating, Refrigerating, and Air-Conditioning Engineers (ASHRAE).

12.5 SUMMARY

With industrial competition becoming more severe than ever, there is less room for breakdowns or production stoppages. A large amount of effort is being put into ensuring the machinery's long-term functionality. Thus, maintenance

Table 12.7 Simulated chiller faults and the simulation methods used for ASHRAE RP-1043 dataset

Sr. No	Fault description	Simulation method
1	Reduced evaporator water flow	% Reduction in the water flow
2	Condenser Fouling	% Blockage in tubes
3	Reduced condenser water flow	% Reduction in the water flow
4	Non-condensable in the refrigerant	Adding nitrogen
5	Refrigerant leak	Reduce refrigerant
6	Refrigerant overcharge	Add refrigerant
7	Excess oil	Adding oil

Source: This dataset is used for developing classification model based on coupling of simulated annealing and deep neural networks (SA-DNN) (Han et al., 2021)

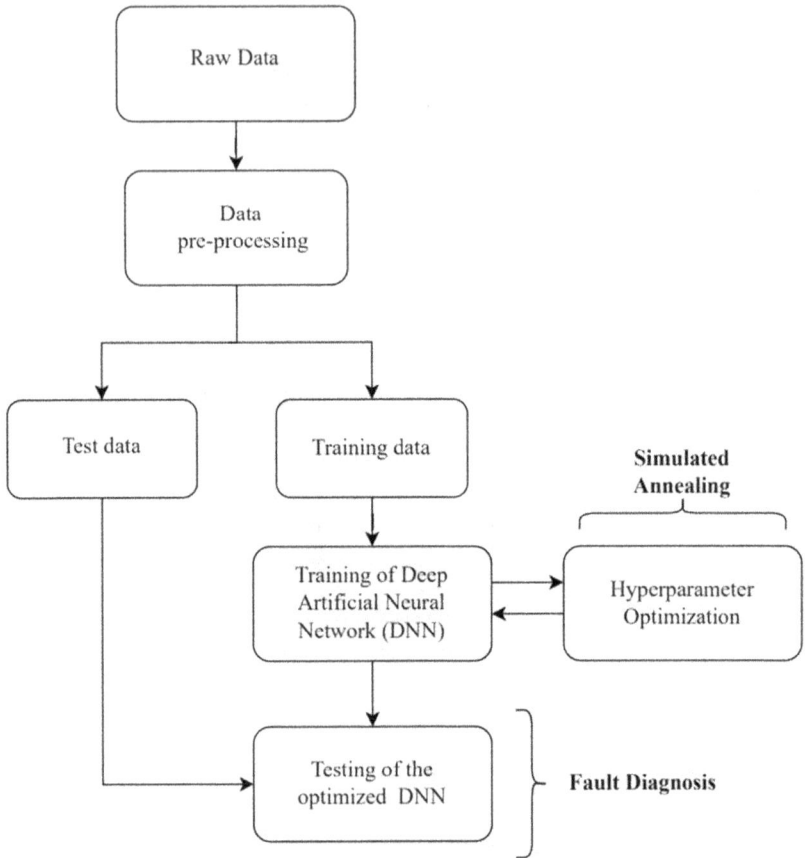

Figure 12.18 SA-DNN-based PdM pipeline (Han et al., 2021).

becomes a key area of focus, and developing an effective and efficient maintenance strategy plays a pivotal role in meeting the productivity requirements. As technology has advanced, we have witnessed a shift in maintenance approaches. In recent years, PdM has attracted researchers and organizations for implementing diagnostics and prognostics concepts. Also, with the availability of advanced sensor technologies, data collection has become more feasible, making ML a core component of the PdM program.

In this chapter, we discussed various ML-based PdM pipelines for engineering systems, focusing on diagnostics and prognostics. In machining systems, we discussed case studies on predicting cutting tools' condition using SVM and fault detection in bearings using two DL methods. For automotive systems, ML-based PdM in gears and brakes for diagnostics and prognostics have been discussed. Later, we have presented the deployment of ML-based PdM for a production line unit, where application of methods like PCA, bagging, and boosting has been explained. Finally, we have also covered

ML-based PdM applications for thermal systems. Here, we have seen examples of fault classification in chillers based on experimental data and boiler faults diagnostics based on simulated data.

The chapter's objective is to provide the readers with the scope of applying ML-based PdM for different engineering systems, to develop an effective diagnostics and prognostics tool.

We believe it motivates the reader to build engineering systems that can utilize the power of PdM and ML, yielding substantial benefits in terms of efficiency and productivity. Interested readers may refer to Dalzochio et al. (2020) and Carvalho et al. (2019) for further reading.

REFERENCES

S. Albawi, T. A. Mohammed, and S. Al-Zawi. Understanding of a convolu- tional neural network. In 2017 international conference on engineering and technology (ICET), pages 1–6. Ieee, 2017.

S. Ayvaz and K. Alpay. Predictive maintenance system for production lines in manufacturing: A machine learning approach using IoT data in real-time. *Expert Systems with Applications*, 173:114598, 2021.

V. Berisha, C. Krantsevich, P. R. Hahn, S. Hahn, G. Dasarathy, P. Turaga, and J. Liss. Digital medicine and the curse of dimensionality. *NPJ Digital Medicine*, 4(1):1–8, 2021.

J. Bhat, U. Bhapkar, and M. Desai. Condition based predictive maintenance of worm gearbox using dr ferrography. Materials Today: Proceedings, 2021.

C. M. Bishop and N. M. Nasrabadi. *Pattern recognition and machine learning*, volume 4. Springer, 2006.

P. Buhlmann. Bagging, boosting and ensemble methods. In *Handbook of com- putational statistics*, pages 985–1022. Springer, 2012.

B. Cai, X. Shao, X. Yuan, Y. Liu, G. Chen, Q. Feng, Y. Liu, and Y. Ren. A novel rul prognosis methodology of multilevel system with cascading failure: Subsea oil and gas transportation systems as a case study. *Ocean Engineering*, 242:110141, 2021.

J. Carrasco, D. López Pretel, I. Aguilera Martos, D.J. García Gil, J. Luengo Martín, F. Herrera Triguero. Anomaly detection in predictive maintenance: A new evaluation framework for temporal unsuper- vised anomaly detection algorithms. *Neurocomputing*, 462:440–452, 2021.

T. P. Carvalho, F. A. Soares, R. Vita, R. D. P. Francisco, J. P. Basto, and S. G. Alcala. A systematic literature review of machine learning methods applied to predictive maintenance. *Computers & Industrial Engineering*, 137:106024, 2019.

D. Chicco and G. Jurman. The advantages of the Matthews correlation coefficient (mcc) over f1 score and accuracy in binary classification evaluation. *BMC Genomics*, 21(1):1–13, 2020.

Z. M. Çinar, A. Abdussalam Nuhu, Q. Zeeshan, O. Korhan, M. Asmael, and B. Safaei. Machine learning in predictive maintenance towards sustainable smart manufacturing in industry 4.0. *Sustainability*, 12(19):8211, 2020.

J. Dalzochio, R. Kunst, E. Pignaton, A. Binotto, S. Sanyal, J. Favilla, and J. Barbosa. Machine learning and reasoning for predictive maintenance in industry 4.0: Current status and challenges. *Computers in Industry*, 123: 103298, 2020.

C. Duan, Z. Li, and F. Liu. Condition-based maintenance for ship pumps subject to competing risks under stochastic maintenance quality. *Ocean Engineering*, 218:108180, 2020.

H. Han, L. Xu, X. Cui, and Y. Fan. Novel chiller fault diagnosis using deep neural network (dnn) with simulated annealing (sa). *International Journal of Refrigeration*, 121:269–278, 2021.

P. E. Hart, D. G. Stork, and R. O. Duda. *Pattern classification*. Wiley Hoboken, 2000.

L. M. Healy. *Logistic regression: An overview*. Eastern Michighan College of Technology, 2006.

M. Hossin and M. N. Sulaiman. A review on evaluation metrics for data classi- fication evaluations. *International Journal of Data Mining & Knowledge Management Process*, 5(2):1, 2015.

R. K. Jha and P. D. Swami. Fault diagnosis and severity analysis of rolling bearings using vibration image texture enhancement and multiclass support vector machines. *Applied Acoustics*, 182:108243, 2021.

R. Jiao, K. Peng, J. Dong, and C. Zhang. Fault monitoring and remaining useful life prediction framework for multiple fault modes in prognostics. *Reliability Engineering & System Safety*, 203:107028, 2020.

P. Kane and A. Andhare. Critical evaluation and comparison of psychoacoustics, acoustics and vibration features for gear fault correlation and classification. *Measurement*, 154:107495, 2020.

R. Khelif, B. Chebel-Morello, and N. Zerhouni. Experience based approach for li-ion batteries rul prediction. *IFAC-PapersOnLine*, 48(3):761–766, 2015.

N. Kolokas, T. Vafeiadis, D. Ioannidis, and D. Tzovaras. A generic fault prog- nostics algorithm for manufacturing industries using unsupervised machine learning classifiers. *Simulation Modelling Practice and Theory*, 103:102109, 2020.

W. J. Lee, H. Wu, H. Yun, H. Kim, M. B. Jun, and J. W. Sutherland. Predictive maintenance of machine tool systems using artificial intelligence techniques applied to machine condition data. *Procedia Cirp*, 80:506–511, 2019.

D. Özgür-Ünlüakın, B. Türkali, A. Karacaörenli, S.Ç. Aksezer. A dbn based reactive maintenance model for a complex system in thermal power plants. *Reliability Engineering & System Safety*, 190:106505, 2019.

L. E. Peterson. K-nearest neighbor. *Scholarpedia*, 4(2):1883, 2009.

H. Pranesh, K. Suresh, S. S. Manian, R. Jegadeeshwaran, G. Sakthivel, and T. A. Manghi. Vibration-based brake health prediction using statistical features–a machine learning framework. Materials Today: Proceedings, 2021.

N. Rout, D. Mishra, and M. K. Mallick. Handling imbalanced data: a survey. In *International proceedings on advances in soft computing, intelligent systems and applications*, pages 431–443. Springer, 2018.

A. Sherstinsky. Fundamentals of recurrent neural network (rnn) and long short- term memory (lstm) network. *Physica D: Nonlinear Phenomena*, 404:132306, 2020.

R. Shohet, M. S. Kandil, Y. Wang, and J. McArthur. Fault detection for non- condensing boilers using simulated building automation system sensor data. *Advanced Engineering Informatics*, 46:101176, 2020.

M. Stender, M. Tiedemann, D. Spieler, D. Schoepflin, N. Hoffmann, and S. Oberst. Deep learning for brake squeal: Brake noise detection, charac- terization and prediction. *Mechanical Systems and Signal Processing*, 149: 107181, 2021.

X. Tian, W. Jiao, T. Liu, L. Ren, and B. Song. Leakage detection of low-pressure gas distribution pipeline system based on linear fitting and extreme learning machine. *International Journal of Pressure Vessels and Piping*, 194:104553, 2021.

L. Wang. *Support vector machines: theory and applications*, volume 177. Springer Science & Business Media, 2005.

Y. Wu, B. Tang, Z. Xiang, H. Qian, J. Mo, and Z. Zhou. Brake squeal of a high-speed train for different friction block configurations. *Applied Acoustics*, 171:107540, 2021.

S. Xiang, Y. Qin, C. Zhu, Y. Wang, and H. Chen. Lstm networks based on attention ordered neurons for gear remaining life prediction. *ISA transactions*, 106:343–354, 2020.

L. Yongbo, D. Xiaoqiang, W. Fangyi, W. Xianzhi, and Y. Huangchao. Rotating machinery fault diagnosis based on convolutional neural network and infrared thermal imaging. *Chinese Journal of Aeronautics*, 33(2):427–438, 2020.

R. Zheng and Y. Zhou. Comparison of three preventive maintenance warranty policies for products deteriorating with age and a time-varying covariate. *Reliability Engineering & System Safety*, 213:107676, 2021. 31

Index

For Product Safety Concerns and Information please contact our EU
representative GPSR@taylorandfrancis.com
Taylor & Francis Verlag GmbH, Kaufingerstraße 24, 80331 München, Germany

9 781032 562971